Techniques of Mushroom Cultivation

The Authors

Dr. Santosh Kumar presently working as an Assistant Professor-cum-Junior Scientist in the Department of Plant Pathology, Bihar Agricultural University, Sabour, Bhagalpur. He did his Ph.D. degree in Plant Pathology from G.B. Pant University of Agriculture and Technology, Pantnagar. He has taught many undergraduate (UG) and postgraduate (PG) courses and actively involved in all teaching activity. He has published many research papers, review articles and book chapters in national and international journal of repute. He has also published one book and 2 extension bulletin in mushroom production. His interest of research includes management of diseases of pulses and rice, mushroom and biocontrol.

Dr. Gireesh Chand is an Associate Professor-cum-Senior Scientist in Department of Plant Pathology, Bihar Agricultural University, Sabour Bhagalpur (Bihar). He has more than decade experience in teaching UG and PG classes, research and extension activities. He authored/co-authored eight books and prepared one practical manual also. He has published 70 research papers, 22 book chapters, 32 popular articles and presented more than 52 research papers in National/International Seminars and Symposia. His interest of research includes molecular plant pathology of crop diseases and their management and mushroom.

Techniques of Mushroom Cultivation

— *Authors* —

Santosh Kumar

Assistant Professor-cum-Jr. Scientist

Gireesh Chand

Associate Professor-cum-Sr. Scientist

Department of Plant Pathology,
Bihar Agricultural University,
Sabour, Bhagalpur – 813 210, Bihar

2018

Daya Publishing House®

A Division of

Astral International Pvt. Ltd.

New Delhi – 110 002

Publisher's Note:

Every possible effort has been made to ensure that the information contained in this book is accurate at the time of going to press, and the publisher and author cannot accept responsibility for any errors or omissions, however caused. No responsibility for loss or damage occasioned to any person acting, or refraining from action, as a result of the material in this publication can be accepted by the editor, the publisher or the author. The Publisher is not associated with any product or vendor mentioned in the book. The contents of this work are intended to further general scientific research, understanding and discussion only. Readers should consult with a specialist where appropriate.

Every effort has been made to trace the owners of copyright material used in this book, if any. The author and the publisher will be grateful for any omission brought to their notice for acknowledgement in the future editions of the book.

Cataloging in Publication Data--DK
Courtesy: D.K. Agencies (P) Ltd. <docinfo@dkagencies.com>

Kumar, Santosh (Assistant professor of plant pathology), **author**.
Techniques of mushroom cultivation / authors, Santosh Kumar, Gireesh Chand.
     pages cm
Includes index.
ISBN: 9789387057814 (International Edition)
1. Mushroom culture. I. Chand, Gireesh, author. II. Title.
LCC SB353.K86 2018 | DDC 635.8 23

Published by   :   **Daya Publishing House®**
                            A Division of
                            Astral International Pvt. Ltd.
                            – ISO 9001:2015 Certified Company –
                            4736/23, Ansari Road, Darya Ganj
                            New Delhi-110 002
                            Ph. 011-43549197, 23278134
                            E-mail: info@astralint.com
                            Website: www.astralint.com

BIHAR AGRICULTURAL UNIVERSITY, SABOUR
BHAGALPUR - 813210 (BIHAR)

Dr. Ajoy Kumar Singh
Vice Chancellor

Phone	: 0641-2452606 (O)
	: 0641-2452605 (R)
Fax	: 0641-2452604
Patna	: 0612-2222267 (O)
Fax	: 0612-2225364
Mob. No.	: +91-8902750693
Email ID	: vcbausabour@gmail.com
URL	: www.bausabour.ac.in

Foreword

Mushroom has taken a prominent place in vegetarian diet. It is becoming popular due to its wider acceptability and rich nutritional and nutraceutical values. Mushrooms require fewer inputs and mostly cultivated using agricultural wastes as compared to growing of agricultural crops. Mushrooms have now become a source of livehlihoods for small and marginal farm families, farmwomen, landless labourers and unemployed youth resulting in economic, nutritional and ecological security. Further, cultivation of mushrooms have become more relevant in this period of decreasing cultivable land due to industrialization, urbanisation and population increase as it requires less space and mostly uses vertical spaces for its production. Due to rising importance of mushroom cultivation the most of the educational institutions have introduced courses related to mushrooms in post-graduate (PG) and undergraduate (UG) programmes.

The book *Techniques of Mushroom Cultivation* written by Dr. Santosh Kumar and Dr. Gireesh Chand combines principal and modern techniques for production of different kinds of edible mushroom. The book has also been designed in pedagogic manner. I feel this book would not only help the students to learn the science and techniques behind mushroom cultivation but will also help to know general information regarding mushroom and its different commercial avenues. It may also be useful to researchers and extension workers to grasp the principle and techniques of different kinds of mushroom production.

I extend my appreciation and good wishes to authors for their efforts for writing this book.

Dr. Ajoy Kumar Singh

Preface

Mushroom has become choice component of the food dish and so its production is surging day by day as being one of the profitable and environmental friendly businesses. Mushroom's production has received significant importance due to their nutritional and medicinal value and presently the cultivation is being carried out in about 100 countries producing 3.4 million tones with annual 6-7 per cent increase. Mushroom has attained the status of a high-tech industry with very high levels of mechanization and automation in developed countries of Europe and America. China leads in mushroom production and alone is reported to grow more than 20 different types of mushroom at commercial scale sharing 46 per cent of the world output. India needs to popularize the mushroom technology through the dissemination of the manual up to the land less farmers, as the production of mushroom does not require land like agricultural crops. The spread of this technology might solve the economic condition of landless and the interested one. Therefore, we have designed this book on *Techniques of Mushroom Cultivation* comprising techniques and practices of spawn production and mushroom cultivation including technological advancements made during the course of time. The methods of isolation, purification and maintenance of the mushroom and disease management strategies have also been dealt. This book will be proved helpful for researchers, as it contains up to date protocols and hands on reproducible and reliable methods of mushroom cultivation. This book will also be helpful for undergraduates, post graduates students, who want to acquire skill in working on spawn production and mushroom cultivation. Every effort has been made by the authors to raise interest of the readers towards the novel practices of mushroom cultivation.

We are highly grateful to Prof. A.K. Singh, Hon'ble Vice-Chancellor, Bihar Agricultural University, Sabour for his consistent encouragement in this regard and to say few words as a preface. Thanks also goes to Dr. R.K. Sohane, Director Extension Education, Dr. J.B. Tomar, Director Research, Dr. Arun Kumar, Dean Agriculture, Dr. B.C. Saha, Dean PGS and Dr. R.R. Singh, Director Seed Research,

BAU, Sabour and Dr. Umesh Singh, Associate Dean-cum-Principal, MBAC, Agwanpur, Saharsa for their support and appreciation for any good in the direction of teaching and research.

We are also thankful to Mr. Tribhuwan Kumar, Assistant Professor-cum-Jr. Scientist, Department of Molecular Biology and Genetic Engineering, BAU, Sabour for his kind assistance in editing of this manual and bringing it into more scientific and lucid manner.

Dr. Ravi Ranjan, Assistant Professor-cum-Jr. Scientist, Department of Molecular Biology and Genetic Engineering, Mr. Deepak Kumar Patel, Assistant Professor-cum-Jr. Scientist, Department of Extension Education, BAU, Sabour and Md. Nadeem Akhtar, Scientist, KVK, Agwanpur, Saharsa also deserve thanks for their helping hand in editing and making it more attractive.

We feel quiet incomplete unless we express our thanks to Dr. Udit Naryan, Former Professor, and S.K. Biswas, Professor, Department of Plant Pathology, CSA University of Agriculture and Technology, Kanpur, UP for taking pain in checking this book so diligently and to hone the book.

I must acknowledged ICAR, New Delhi and RKVY, Bihar Govt., Bihar for their financial support to run the research project, as working on this project enhanced my intellectual property and resulted in the motivation to write this book.

Finally we owe the obvious support and sacrifice of our families for moving every single step of success. Of course, it would be unrealistic to suppose that the text is free from errors and notification in this regard and suggestions for the rectification and improvement of this book will be greatly appreciated.

Dr. Santosh Kumar

Dr. Gireesh Chand

Contents

Foreword   *v*

Preface   *vii*

List of Abbreviations   *xi*

1. Essential Laboratory Practices and Principle and Operation of Equipments   1

2. Introduction   13

3. General Morphology of Mushroom   25

4. Life Cycle and Strain Improvement (Breeding Programme) of Mushroom   29

5. Identification of Mushroom   33

6. Preparation of Culture Media, Preservation and Storage of Mushroom Cultures   37

7. Techniques of Pure Culture of Mushroom   43

8. Techniques of Spawn Production   49

9. Techniques of Compost and Casing Preparation   57

10. Techniques of Spawning   63

11. Cultivation Technique of Button Mushroom (*Agaricus* sp.)   67

12. Cultivation Technique of Oyster Mushroom (*Pleurotus* sp.)   73

13. Cultivation Technique of White Milky Mushroom (*Calocybe indica*)   79

14. Cultivation Technique of Paddy Straw Mushroom
(*Volvariella volvacea*) 83

15. Cultivation Technique of Other Edible Mushrooms 89

16. Cultivation Technique of Medicinal Mushrooms 93

17. Poisonous Mushroom 101

18. Post Harvest Management of Mushroom 105

19. Mushroom Cooking and its Value Addition 111

20. Major Diseases, Pests, Disorders of Mushroom and their Management 117

21. Ethics and Economics in Mushroom Cultivation 127

22. Marketing, Export and Financial Services in Mushroom Business 133

23. Mushroom Cultivation Management 137

Glossary 145

Auxiliary Information 151

Appendix 153

Index 159

List of Abbreviations

AI: Active ingredient

AICRP: All India Coordinated Research Project on Mushroom

BOD: Biological Oxygen Demand

$CaCO_3$: Calcium carbonate

CAP: Controlled Atmosphere Packaging

$CaSO_4$: Calcium Sulphate

DMR: Directorate of Mushroom Research

HEPA: High Efficiency Particular Air

ICAR: Indian Council of Agricultural Research

ISMS: International Society of Mushroom Science

KMS: Potassium Meta Bisulphite

MAP: Modified Atmosphere Packaging

MEA: Malt Extract Agar

MSI: Mushroom Society of India

NCMRT: National Centre for Mushroom research and Training

NRCM: National Research Centre for Mushroom

PDA: Potato Dextrose Agar

PP: Poly Propylene

RH: Relative Humidity

SMS: Spent Mushroom Substrate

TSS: Total Soluble Salt

UV: Ultra Violet

WHO: World Health Organization

Chapter 1

Essential Laboratory Practices and Principle and Operation of Equipments

Essential Laboratory Practices

Proper basic knowledge, precautions, working principles of the equipments and information regarding nature of the chemicals are prerequisites for commencing any laboratory work. Spawn laboratory and mushroom house/cropping room are two essential subunits of mushroom production unit. Proper maintenance of the mushroom production unit and knowing of essential laboratory practices determine the success and production of mushroom, which are being dealt here.

A. Spawn Laboratory (Inoculation and Incubation Chamber)

☆ The lab environment must be calm and tidy.

☆ Work benches must be cleaned before starting and after finishing of work.

☆ Chemicals, glassware and other articles in the lab should be placed at proper and classified places.

☆ Eating, drinking and talking must be avoided while working.

☆ All the electric supplies must be plugged out if there is no use.

☆ Proper disposal system must be available to dispose waste.

☆ Water knob should be closed tightly after use.

☆ Work must be planned in before, in order to finish all operations at predetermined time.

☆ The precautions and care taken must be known before operation of any equipment.

B. Mushroom House/Cropping Room

☆ Old and infected bag should be discarded immediately form the cropping room.

☆ Mushroom house/cropping room should be inspected regularly and contaminated one must be discarded immediately.

☆ Hygienic sanitation are the most important operation under cropping room.

☆ Working person should be clean and dirt free.

☆ Worker should wear a laboratory coat, a mask, a lab slipper and a pair of hand gloves while working.

☆ Entry and exit of visitors and even of the workers should be as minimum as possible.

☆ Filling bag must be tagged and labelled with species name, quantity, date of spawning.

☆ Recording of activities of mushroom house should be noted in the record book.

General Cleaning Chemicals in the Laboratory

Petri dishes, culture tubes, flasks, beakers, measuring cylinders, forceps, spatula, cork borer, slides *etc.* are the common elements which are used in the mushroom culture. Before starting any work glassware are dipped in 1 per cent HCL solution overnight and washed under running tap water and finally rinsed with distilled water. Acids ethanol and dichromate solution may also be used for cleaning purposes. Cleaning of such glassware of general use in the mushroom spawn laboratory is essential to avoid contamination of microorganisms or any germs.

General Equipments, their Uses and Handling

Autoclave

Principle

High temperature, pressure and longer duration of these variable kills even thermophillic microorganism. Moist heat generated in the autoclave destroys microorganism through irreversible denaturation of enzyme and structural protein. Pressure cooker may be used, if autoclave goes out of order, as the principle of autoclave and pressure cooker is same.

General Description

The Autoclave is used for sterilization of various utilities like culture media, grain substrates, under saturated steam pressure at any selected point between 10 to 20 pounds per square inch (p.s.i.). Generally the steam sterilization is done at 121°C at 15 p.s.i. for 15 minutes. However, the time is counted after the pressure reaches 15 p.s.i. Autoclave may be either vertical or horizontal cylindrical double or triple walled mounted on a strong stand. The inner chamber (container) is made up of stainless steel and outer surface is of painted mild steel. Following are the temperatures at various pressures:

Temperature (ºC)	p.s.i.
107	5
110	7
115	10
121	15
126	20

Precautions

The level of water should be checked before operation. All exhaust vents and safety valves and chamber should be kept clean. Over sterilization of any material should be avoided otherwise it may cause hydrolysis of the biochemical compounds. Heat sensitive articles should not be sterilized under autoclave.

Figure 1.1: Autoclave.

Hot Air Oven

Principle

It is based on principle of dry heat sterilization. It kills the microbes by oxidizing their chemical constituents mainly cell constituents. This devise is commonly used for sterilization of many objects such as glass wares like Petri dish, pipettes, flasks and other usable under dry heat during many biological exercises. It is a vertical steel

Figure 1.2: Hot Air Oven.

box with double or triple walled body of aluminium of stainless steel partitioned with wire mesh trays inside. The body of the device is provided with heating element between the walls either at the bottom of the box (bottom heated) or on all three side of the body (universal heating). The outside of the body is painted with an epoxy powder coating. Some ovens are also provided with system for the circulation of hot air in between the inner chamber and insulation through forced air moved by motorized blower so as to minimize the temperature variation at any point in working space. Ovens are available in varying capacity. A thermostat control is provided to maintain the temperature inside with the sensitivity less than ±3°C. The door is provided with synthetic rubber gasket to make it air tight. The range of temperature inside varies from 50 to 300°C or more. The front of the oven is provided with a digital temperature controller- cum- indicator and power switch on/off. In ovens, the temperature is maintained above the ambient temperature.

Generally the dry heat sterilization is done at 160°C for 1 hours However, the temperature required for sterilization may vary with the duration of exposure, as mentioned below:

Temperature (ºC)	Required time
120	8 hours
140	3 hours
160	1 hours
180	20 minutes

Precautions

Do not open the device immediately after the process is finished. It would cause contamination in the material and the glasswares, after exposure with cool air on immediate opening the oven. The glass materials should be wiped and dried before keeping inside the chamber, otherwise it may break. The air, within the oven, should be circulated by a fan to ensure required temperature to every part and the article should be placed properly so as not to impede flow of air.

BOD Incubator

The biological oxygen demand (BOD) incubator maintains a range of temperature below and above the ambient temperatures required for growth and multiplication of mushroom fungus. It is a vertical steel chamber shaped as an almirah made up of double walled body. Incubators are available in varying capacities. Temperature inside incubator may be maintained from 5 to 50ºC with an accuracy of 1ºC. The required temperature is adjusted in the incubators as per the information required for growth and multiplication of the mushroom. Incubator is provided with both heating and cooling systems. Heating may be of two types. Bottom heating and universal heating (in which the heating element is placed in all three side walls) with a thermostatic control while cooling is maintained by compressors. It is provided with air circulation fans for uniform distribution of temperature inside.

Precautions

The door of the incubator should be opened as per need only otherwise it may causes variations in the maintained temperature. Incubators are cleaned and sterilized at frequent intervals to avoid contamination of the materials.

Laminar Flow Cabinet

Laminar flow machine provides an aseptic or micro-organism free environment for performing various activities such as pouring of sterilized media in sterilized Petri dish, isolation, inoculation and transfer of mushroom fungus during application of different methodologies, which require aseptic or sterilized environment. The cabinet is fitted with both pre filter and high efficiency particular air (HEPA) filter. Air is drawn through pre filter and is made to pass through highly effective HEPA. The HEPA filter removes nearly all microorganism and particles as small as 0.3 micron to 0.45 micron having high efficiency as 99.99 per cent. The working area is illuminated by fluorescent light and UV light at the top. UV light is switched on 15 minutes before working to remove/kill the microorganism, if any present in the

Figure 1.3: Biological Oxygen Demand (BOD) Incubator.

area of working space. Surface of the laminar flow is cleaned with cotton soaked with sprit containing 70 per cent alcohol before working.

Precautions

☆ Proper care is taken to not expose any part of the body before the UV light as the exposure may be carcinogenic or mutagenic.

☆ Jewellery from the hands and wrists must be removed from the hand followed by washing and cleaning.

☆ Objects must be put in a manner to get full benefit of the airflow of cabinet.

☆ After every use the surface of the laminar flow must be cleaned with sprit containing 70 per cent alcohol.

☆ Blower must be kept on, while working.

Figure 1.4: Laminar Air Flow.

Refrigerator/Deep Freezer

It is used to preserve and to maintain the culture in pure form for further use. Generally refrigeration is carried out at low temperature 0-5°C. At this temperature

Figure 1.5: Deep Freezer.

all metabolic processes gets slow down and culture may be maintained without losing their identity for a longer period. Deep freezer has all specifications similar to that of refrigerator except that the temperature maintained there is below 0°C.

Weighing Balance

It is used to weigh chemical and other ingredients required during the course of experiments. Various types of balance such as single pan, analytical, electrical balance *etc.* is available. The accuracy of weighing is determined by the sensitivity of the balance. Balance is calibrated by internal calibration before use. Weighing boat is placed and tared zero. Electronic balance is more accurate, sensitive and is easy to handle.

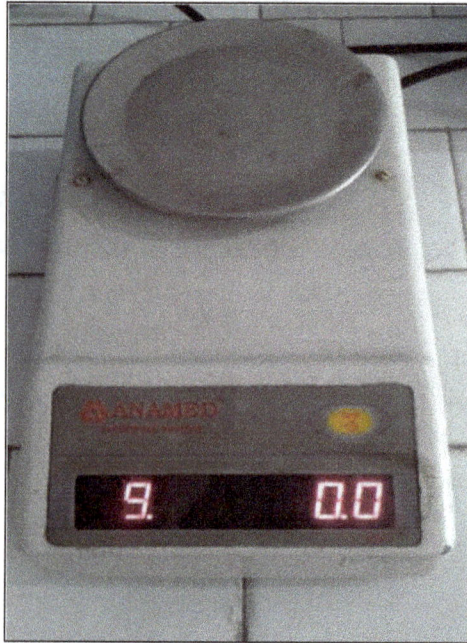

Figure 1.6: Electronic Balance.

Precautions

Electronic balance is highly sensitive and even air of fans may affect the accuracy of the measurements. After every use, pan must be cleaned with dry tissue paper and kept in dust proof chamber.

Distillation Plant

This is required to produce distilled water, used in media preparation and other experiments.

Shaker

A lab shaker is used to agitate/shake liquids materials within lab containers for procedures such as assays, culturing, staining *etc.* Most lab shakers can operate in incubators, refrigerators, and fume hoods. Various types of lab shaker are available, each offering a different shaking pattern. An orbital shaker moves its platform in a circular motion; a reciprocating shaker moves its platform from side to side; a rocking shaker tilts its platform up and down; a vortex shaker spins its platform or cup head; a vibration shaker has a vibrating platform; and a wrist-action shaker replicates hand mixing. A shaker is designed to hold microplates, and is available in various shaking patterns.

Figure 1.7: Shaker.

Hot Plate

A hot plate is a portable self-contained tabletop small appliance that features one, two or more electric heating elements or gas burners and are generally used to heat glassware or its contents. Some hot plates also contain a magnetic stirrer, allowing the heated liquid to be stirred automatically. Hot plate does not only reduce the temperature of the glass, but also slows down the rate of heat exchange and encourages even heating. This works well for low boiling point operations or when minimum temperature of heat source is high.

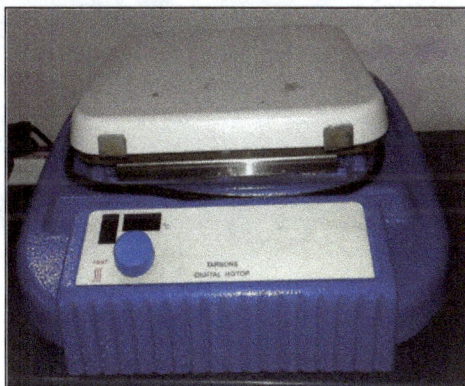

Figure 1.8: Hot Plate.

Petri Dish and Test Tube

A Petri dish/Petri plate or cell-culture dish is used to put the culture media inside for mycelial growth of mushroom. Petri dishes are often used to make agar plates for microbiological studies also. A test tube/culture tube widely used to hold, mix, or heat small quantities of solid or liquid chemicals, especially for qualitative experiments and assays. Culture tubes are often used for slow growth of culture and its preservation.

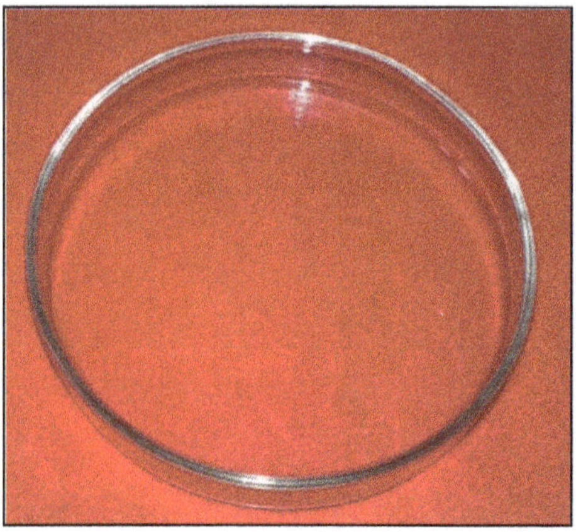

| Figure 1.9: Petri Plate. | Figure 1.10: Test Tube. |

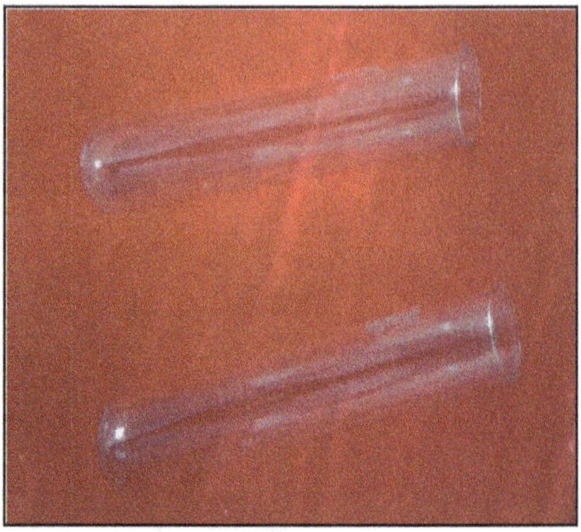

Spirit Lamp

It is lamp which is burnt with the help of spirit (70 per cent alcohol) and to provide aseptic environment around its flame and transferring the culture media into the culture plate inside laminar flow.

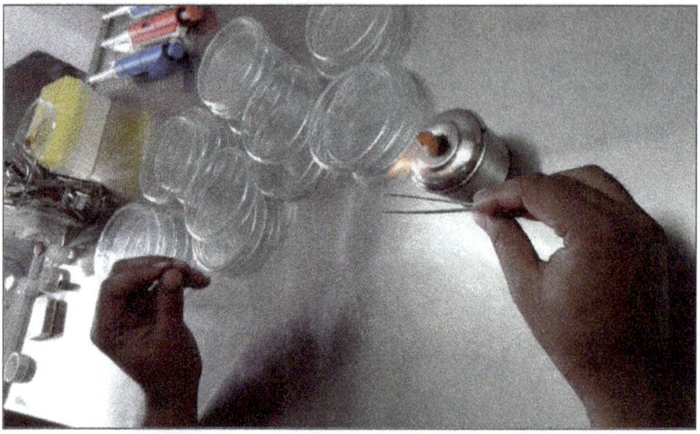

Figure 1.11: Flame Sterilization with the Help of Spirit Lamp.

Hygrometer and Thermometer

Hygrometer is used to measure relative humidity in spawn laboratory as well as in mushroom house/cropping room. Thermometer is a device used to measure required temperature for proper spawn growth and fruiting of mushroom.

Figure 1.12: Hygrometer.

Humidifier

It is an important device to maintain desired level of humidity (moisture) for proper growth of spawn and fruiting of mushroom. Some time sudden changes in room humidity takes place, therefore it becomes essential to have a device which

Figure 1.13: Humidifier Set.

can restore the humidity in a mushroom house to a required level. There are many types of humidifiers depending upon the need and requirements.

Chemicals Used during Spawn Production and Mushroom Cultivation

Calcium Carbonate ($CaCO_3$) and Calcium Sulphate ($CaSO_4$)

0.5 per cent $CaCO_3$ and 2 per cent $CaSO_4$ are required for the prepartion of spawn. Both chemicals are added and mixed with grain substrate just before the grain filling in glucose bottle or polypropylene bags (pp bag). Calcium carbonate is used to avoid sectoring while calcium sulphate to avoid clumping of grain between them.

Bavistin and Formalin

Bavistin (fungicide) and formalin (gaseous form) are used for the sterilization/ treatment of straw substrates, compost, casing soil and making it fungi, bacteria and other microbes free.

Method of Sterilization of Various Objects

Sterilization is a procedure used for complete destruction or elimination of all forms of life or living organism from the object due to denaturation of enzyme. Selection of the sterilization methods depend on desired efficiency, its applicability, easy to use, availability and effect on properties of the object to be sterilized.

1. **Dry heat sterilization:** It is done by using hot air oven for the sterilization of glasswares such as Petri dishes, flasks, pipettes *etc*. Generally the dry heat sterilization is done at 160°C for 1 hours and 180° C for 30 minutes.

2. **Moist heat sterilization:** It is done through autoclave for the sterilization of culture media, grain substrate and generally the moist heat sterilization is done at 121°C at 15 p.s.i. for 15 minutes.

3. **Flame sterilization:** Flame is done for sterilization of metal objects such as needle, tips of forceps, lips of flasks, culture tube, the slides and cover slips *etc*. It is done by objects dipped in 70 per cent alcohol or rectified spirit followed by flaming.

4. **Chemical sterilization:** Chemicals such as mercuric chloride (0.1 per cent), sodium hypochloride (1 per cent), alcohol (70 per cent ethanol) and formaldehyde (4 per cent) are used for surface sterilization of mushroom tissues.

Chapter 2

Introduction

Mushroom

Mushrooms are a fleshy, macroscopic, saprophytic, spore-bearing fruiting structure of a fungus (edible fungus) belonging to class Basidiomycetes except few which belong to Ascomycetes. It grows on dead and decaying organic materials absorbing nutrients from decaying substrates using fine, usually thread like structures (mycelium) which penetrate into the substratum. Mycelium absorbs sufficient food materials and grows profusely forming reproductive structure and further fruiting body, commonly known as mushroom.

Mushroom mycelium secretes extracellular enzymes during vegetative phase which break down compounds such as cellulose and lignin present in the substrate. The degraded compounds are then absorbed by the hyphae and the mycelium enlarges laterally and sometimes several meters in diameter in the substrate.

Types of Mushroom

They are grouped into four categories.

1. **Edible mushroom:** These are fleshy and edible *i.e. Agaricus* sp., *Pleurotus* sp., *Calocybe indica, Volvariella* sp. etc.

2 **Medicinal mushroom:** These are considered to have medicinal applications *i.e. Lentinus edodes, Ganoderma lucidum, Cariolus* and *Shyzophyllum*.

3 **Non edible or poisonous mushroom:** These are proven to be, or suspected of being poisonous. *i.e. Amanita phalloides, A. muscaria, A. verna* etc.

4. **Miscellaneous category:** It includes a large number of mushrooms which properties are poorly defined.

Note

Agaricus bisporus is the most popular, more economic species and is extensively cultivated across the globe. However, fruiting of this species requires low

**Table 2.1: Important Species of Edible Mushroom and
Year-wise Production Schedule**

Sl.No.	Scientific Name	Common Name	Production Month	Temperature
1.	*Agaricus bisporus*	Button mushroom or Khumb mushroom	Mid November to mid February	18-20ºC
2.	*Pleurotus* sp.	Dhingree mushroom	October-March	22-25ºC
3.	*Calocybe indica*	Milky mushroom or Dudhiya mushroom	June-August	30-35ºC
4.	*Volvariella volvacea*	Paddy straw mushroom/ Chinese mushroom	September to November	30-32ºC
5.	*Lentinula edodes*	Shitake	February to May	14-22ºC
6.	*Auricularia* sp.	Black ear	June-August	18-34ºC

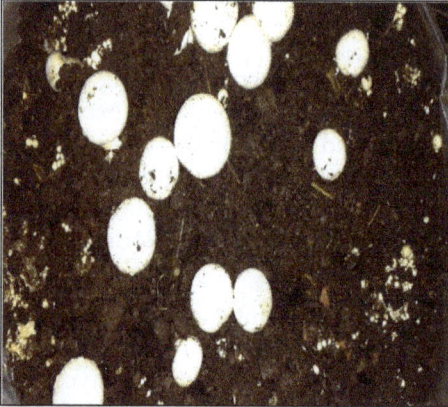

Agaricus bisporus

Pleurotus sp.

Calocybe indica

Volvariella volvacea

Figure 2.1: Major Cultivated Species of Edible Mushroom.

temperature; its cultivation is restricted to the cold climatic areas. Button mushroom is widely grown throughout the world and holds first rank in production.

Choosing the Mushroom Species

☆ Prevailing climatic condition.

☆ Location of mushroom farm.

☆ Availability of the substrate for chosen species.

☆ Available technology for chosen species.

☆ Acceptability to the local people on to the market for which it is intended.

History of Mushroom Cultivation in India

The history of mushroom cultivation goes back to the *Vedic* period. Classical religious scriptures like 'Rig Veda' and 'Atharva Veda' also mention the use of juice from fly agaric mushroom (*Amanita muscaria*) as an intoxicating drink, named as "Soma". However, the systematic research on mushroom domestication did not persist for longer period. Practices of edible mushrooms cultivation in India has originated of late, although methods of cultivation for some species were known for years. The important historical developments made in the cultivation of edible mushrooms in India is given in Table 2.2.

Table 2.2: Historical Developments of Mushroom in Chronological Order

Year	Name of Contributors
1886	N.W. Newton exhibited some edible mushrooms in the annual flower show of Horticultural Society of India
1896-1897	Dr. B.C. Roy of the Calcutta Medical College carried out chemical analysis of the local mushrooms prevalent in caves or mines.
1908	Sir David Prain searched edible mushrooms from various parts of India
1918	Kirtikar observed the occurrence of mushrooms from Calcutta
1921-26	S.R. Bose successfully raised two *Agaricus* on sterilised dung medium, details of which were published in the Indian Science Congress held at Nagpur during 1926.
1939-40	Attempts on experimental cultivation of paddy straw mushroom (*Volvariella*) were first undertaken by the Department of Agriculture, Madras (T.N.).
1940	Su and Seth described the procedure for spawn and cultivation of *Volvariella*
1940	Bose and Bose discussed some methods for growing mushrooms on horse manure
1941	Padwick reported successful cultivation of *Agaricus bisporus* from various countries but without much success in India.
1943	Thomas *et al.*, first developed successful cultivation techniques for paddy straw mushroom, *Volvariella volvacea* in Agricultural College, Coimbatore. This lead to the spread of cultivation of this mushroom in all parts of India.
1947	R.P. Asthana reported better yield of paddy straw mushroom by adding red powdered dal to the beds. He suggested April-June as the most suitable period for culturing this mushroom in Central Provinces and also carried out the chemical analysis of this mushroom.

Contd...

Table 2.2–*Contd...*

Year	Name of Contributors
1961	First cultivation of *Agaricus bisporus* scheme entitled "Development of mushroom cultivation in "Himachal Pradesh" was started at Solan by the H.P. Govt. in collaboration with ICAR.
1962	Bano *et al.*, obtained increased yields of *Pleurotus flabellatus* on paddy straw
1964	Cultivation of *Agaricus bisporus* on experimental basis was started by CSIR and state Govt. at Srinagar in Jammu and Kashmir.
1965	E.F.K. Mantal, an FAO Expert guided construction of growing facilities and research on spawn production, synthetic compost.
1971	ICAR sponsored coordinated scheme on mushroom research was started at Solan, Ludhiana, Bangalore and New Delhi
1974	Dr. W.A. Hayes an F.A.O., Mushroom Expert guided in further improving the method of compost preparation, pasteurization and management of important parameters in the mushroom house. New compost formulations, casing materials and important parameters like nitrogen content in the compost, moisture in the casing soil, air movement and maintenance of proper environmental factors were also standardized which raised the mushroom yields from 7 to 14 kg/m^2.
1977	A 1.27 crore, Mushroom Development Project was lunched under the U.N.D.P. by the Deptartment of Horticulture (H.P.) wherein the services of Mr. James Tunney were made available. He got a bulk pasteurization chamber constructed and made available ready compost and casing soil to the growers of H.P. the UNDP Project was concluded during 1982 and since then Department of Horticulture (H.P.) is running the project.
1982	The Indian Council of Agriculture Research (ICAR) sanctioned the creation of National Centre for Mushroom Research and Training (NCMRT) during VI Plan on Oct. 23, 1982 with the objectives of conducting research on problems of mushroom production, preservation and utilization and to impart training to scientists, teachers, extension workers and interested growers. NCMRT started functioning w.e.f. 1983.
1983	The All India Co-ordinated Mushroom Improvement Project (AICMIP) was sanctioned by Indian Council of Agriculture Research (ICAR) with its headquarters at the National Research Centre for Mushroom, Solan (now DMR) and Directorate as the Project Coordinator during the VI Plan w.e.f. 1st April, 1983. The six centres initially sanctioned were located at G.B. Pant University of Agric. and Tech., Pantnagar (Uttarakhand); P.A.U. Ludhiana (Punjab); TNAU, Coimbatore (TN); B.C.K.V.V., Kalyani (West Bangal); MPAU, College of Agriculture, Pune (Maharashtra) and C.S. Azad Univ. of Agric. and Tech., Kanpur (Uttar Pradesh). In subsequent years Kanpur and Kalyani centres have been deleted and IGKVV, Raipur (Chhattisgarh) and NDUA&T, Faizabad (Uttar Pradesh) were added. Presently, 23 co-ordinating centres and 9 co-operating centres are functioning under AICRP–Mushroom, which are located at different states Agricultural Universities and ICAR institutes.
1997	NCMRT was renamed as National Research Centre for Mushroom (NRCM)
2008	NRCM upgraded to Directorate of Mushroom Research (DMR) in December, 2008

The research on this mushroom got momentum after establishment of AICRP-Mushroom at Solan. India has registered twenty-fold increase in production of mushrooms during the last four decades, among them button mushroom continues to occupy a prominent place contributing about 85 per cent of the total mushroom production.

Directorate of Mushroom Research

An ICAR sponsored coordinated scheme on mushroom research was started at Solan, Ludhiana, Bangalore and New Delhi in 1971. Thereafter, a pilot scale cultivation plan for button mushrooms was started in Himachal Pradesh during 1974 under UNDP. E.F.K. Mantal, one of the FAO experts guided the Department of Agriculture, Himachal Pradesh for the construction of a modern spawn laboratory. His efforts paved the way for the introduction of button mushroom in India. Meanwhile the National Centre for Mushroom Research and Training (NCMRT) was established on 23rd October, 1982 at Solan, Himachal Pradesh. This centre became functional with effect from 8th June, 1983 under the aegis of Indian Council of Agricultural Research, New Delhi, which was renamed as National Research Centre for Mushroom in 1997 and was upgraded to Directorate of Mushroom Research (DMR) in December, 2008. Solan (Himachal Pradesh) is known for mushroom cultivation widely and entitled as "Mushroom City of India". This Directorate is the only organization, which is exclusively dedicated to mushroom research and development in the country. It has developed array of technologies for cultivation of different variety of mushrooms as per concerned the agro-climatic regions of the country.

Current Status of Mushroom Production in India

Agaricus bisporus accounted for over 90 per cent of total mushroom production value while *Pleurotus, Volvariella, Lentinula, Flammulina, Hypsizygus, Hericium, Morchella,* and *Grifola* were the main specialty genera cultivated in world. India has varied agro-climate zones with abundance of agricultural residues and plenty of manpower making it suitable for cultivation of different variety of mushrooms. Major part of agricultural wastes in India is either let out to decompose naturally or burnt *in situ*. At present, four mushrooms varieties *viz.*, button mushroom (*Agaricus bisporus*), oyster mushroom (*Pleurotus* sp.), paddy straw mushroom (*Volvariella* sp.) and milky mushroom (*Calocybe indica*) have been recommended for around the year cultivation for edible purposes.

Commercial cultivation of mushrooms had become possible with the joint effort of scientists and farmers. Annual mushroom production has increased upto 80,000 ton in the year 2006 as compared to that of mere 1,000 ton in 1981. Marginal and small production units have fifty per cent contribution in this production and the rest goes to the credits of industrial establishments. Mushroom husbandry has become now, the major sources of income for farmers of many states like Punjab, Uttarakhand, Haryana, Uttar Pradesh, Tamil Nadu, Himachal Pradesh, Orissa, Andhra Pradesh, Maharashtra, Kerala, and North eastern regions of India. First three major producers of mushrooms in India are Punjab (35,000 MT) Tamil Nadu, (15,000MT) and Andhra Pradesh (5000MT). Button mushroom (*Agaricus bisporus*) constitutes about 90 per cent of total production in India while other cultivated mushrooms *viz., Pleurotus, Volvariella, Lentinula, Auricularia* and *Calocybe* are very marginal.

Importance of Mushroom

Mushroom has been part of our human diet since the time immemorial. It was used as food even before man could realize the use of other organisms. Undoubtedly, mushroom was one of the earliest easily available foods for human beings, and now a days it has turned into an exotic and luxurious food reserved for the people. Mushrooms are food materials for both the rich and poor. It is noteworthy that not all mushrooms are edible. It is believed that less than ten per cent of mushrooms varieties are edible, and the rest are poisonous. Only around 3000 varieties are edible out of 14,000 known varieties. Edible mushrooms are often described as vegetables or herbs, but these are actually fungi. People consume the edible ones for the exotic yet subtle flavour, the nutritional benefits as well as for the medicinal value. It can be found in markets, as fresh, packed or dried. The health benefits of edible mushroom have it choice food material for fitness enthusiasts and those interested in losing weight.

Uses of Edible Mushroom

☆ The edibility of any variety of mushrooms is determined by safety and taste, as many of the non-edible varieties can actually be poisonous.

☆ Mushrooms are relished because of their mild flavour and distinctive texture.

☆ Edible mushrooms can be used in a variety of dishes because they can impart flavour to the meal or even take on the flavour of other ingredients.

☆ Mushrooms are often added to soups, salads, pasta and sandwiches. Edible mushrooms have become increasingly popular for their health benefits. A quick look at some edible mushroom nutrition info will explain why mushrooms have come to be associated with health foods.

☆ Edible mushrooms are generally cooked into soups, stews, and stir-fries; however, some mushrooms are eaten raw. Some of the mushrooms that are eaten raw include porcinis, portobellos, creminis and some russulas.

Nutrition Facts in Edible Mushroom

☆ Nutritional value of edible mushroom is off course dependent on the method of preparation, so if you're looking to prepare a healthy meal avoid using other ingredients that would increase the caloric value of the meal.

☆ Avoid frying or using fatty and oily food dressings.

☆ In addition to their low caloric value and low sodium and fat content there are several essential nutrients in edible mushrooms that add to their value as a health food. The copper content in mushrooms is also considerable. Edible mushrooms are rich in proteins as well as other nutrients like iron, potassium, selenium and some B complex vitamins.

Precaution before Using Edible Mushrooms

No matter how healthy and nutritious mushrooms may be, you need to exercise a great deal of caution when consuming them. Only by commercially sold mushrooms that are marketed by reputed vendors, as mushrooms can easily be contaminated and some varieties of poisonous mushrooms can easily be confused with edible varieties.

A. Nutritional Importance (Used as Vegetable)

The popularity of mushrooms is based not on the nutrients that they contain but mostly on the exotic taste and their culinary properties, whether eaten alone or in combination with other food. It is well known that mushroom is full of nutrients and can therefore make a very important contribution to human nutrition. Protein is one of the most important nutrients in food, being particularly important for building body tissues. Mushrooms with protein content ranging from 3-7 per cent when fresh to 25-40 per cent when dry can play an important role in enriching human diets when meat sources are limited. The protein content in the mushroom is almost equal to that of corn, milk and legumes, although still lower than meat, fish and eggs. Mushroom also contains all the essential amino acids as well as the commonly occurring non-essential amino acids and amides. Lysine, which is low in most cereals, is the most important amino acid in mushroom. Mushroom also rank quite high in their vitamin content, which includes significant amounts of vitamin C. Although devoid of vitamin A, mushroom makes up for that with their high riboflavin, thiamine and cyanocobalamin (Vit. B_{12}) content, the latter usually being found only in animal products. Mushroom are also good source of minerals such as calcium, potassium, sodium and phosphorus in addition to folic acid, an ingredient known for enriching the bloodstream and preventing deficiencies. Iron is also present in an appreciable amount in mushroom and together with phosphorus, can provide a good proportion of the recommended daily dietary needs.

☆ Mushroom contains quality crude protein and essential amino acid such as lysine and tryptophan, fibre, carbohydrate, vitamins (especially vitamin B, C, niacin, ergothionine).

☆ Mushroom are low calorific, fat, starch, cholesterol free and gluten free, and has very low quantity of sodium, yet contain ions such as potassium, selenium.

Table 2.3: Nutrition Value in Major Edible Mushroom in per cent/100 mg Dry Weight

Mushroom Species	Protein (Per cent)	Fat (Per cent)	Fibre (Per cent)	Ash (Per cent)	CHO (Per cent)	Energy (kcal/kg)
Agaricus bisporus	28.1	8.9	8.3	9.4	59.4	353
Pleurotus sp.	30.4	2.2	8.7	9.8	57.6	345
Calocybe indica	17.7	4.1	3.4	7.4	64.3	363
Volvariella volvacea	29.5	5.7	10.4	9.8	60.0	374
Lentinus edodus	32.9	3.7	28.9	9.6	47.6	356

Table 2.4: Minerals Value in Major Edible Mushroom in per cent/100 mg Dry Weight

Mushroom Species	Calcium (Per cent)	Phosphorus (Per cent)	Iron (Per cent)	Sodium (Per cent)	Potassium (Per cent)
Agaricus bisporus	23	1429	88	106	4762
Pleurotus sp.	71	677	17.1	374	3455
Volvariella volvacea	98	476	8.5	61	3793
Lentinus edodus	23	1257	5.5	18	2700

Table 2.5: Minerals Value in Major Edible Mushroom in per cent/100 mg Dry Weight

Mushroom Species	Thiamine (Per cent)	Riboflavin (Per cent)	Niacin (Per cent)
Agaricus bisporus	1.1	5.0	55.7
Pleurotus florida	0.35	2.97	64.88
Pleurotus sajorcaju	1.16 - 4.8	-	46.108
Volvariella volvacea	0.32	1.63	47.55
Lentinus edodus	7.8	4.9	54.9

B. Medicinal Importance

Mushroom has cardio-vascular, chemopreventive, antitumor, anticancer, antiviral, antibacterial, antiparasitic, anti-inflammatory, and antioxidative, antidiabetic, anti-herpetic, anti-aging and also anti-ulcer properties, which have been dealt briefly.

1. Diabetic Patients

Mushroom is preferred by diabetic and hypertension patients due to low sodium content, low caloric value, low sugar and low cholesterol free fat quantity. For *e.g. Pleurotus* spp., *Ganoderma lucidum*.

2. Hypertension

This also makes it ideal for any diet formulated for patients suffering from hypertension. Mushrooms are rich in potassium, which enhances preferences in a hypertension diet. For *e.g. Pleurotus* sp.

3. Anaemia

Mushroom helps to reduce anaemia due to presence of folic acid. For *e.g. Pleurotus* sp.

4. Reduce High Blood Pressure

Mushroom helps to reduce high blood owing to the presence of potassium, an essential mineral. For *e.g. Ganoderma lucidum*.

5. Antioxidant property

Mushrooms have antioxidant property due to presence of ergothioneine. For *e.g. Pleurotus* sp. *Ganoderma lucidum*.

6. Prevention of Cardio-vascular Diseases

The high proteins, sterols, macro-elements and low calorific value make mushroom choice food for prevention of cardio-vascular diseases.

7. Immune System Enhancer

Mushroom helps to raise immunity in the human body. For *e.g. Ganoderma lucidium.*

8. Hormone Stimulator

Few species of mushroom stimulates hormone secretion *e.g. Cordyceps sinensis.*

9. Weight Loss

Mushrooms are extremely low in calorific value besides containing negligible amounts of sodium and fat. This alone makes mushrooms, a viable food choice for one who wants to lose weight.

10. Cell Damage

Edible mushrooms also contain significant amounts of riboflavin and niacin, and are rich in selenium, an antioxidant that helps to protect cell damage caused by free radicals. It also stimulates non-specific immune system, reduces blood cholesterol and glucose levels.

C. Nutraceuticals and Dietary Supplements

The mushroom nutraceutical is a refined/partially defined mushroom extractive which is consumed as capsules or tablets as a dietary supplement (not a food) and potential therapeutic ingredients. Regular intake may enhance the immune responses, increasing resistance to disease and, in some cases, causing regression in disease proneness.

D. Agro-waste Degradation Potential

Mushroom produces industrially useful enzymes, which have potential to degrade waste materials from agriculture and industry.

E. Spent Mushroom Substrate (SMS)/Spent C-ompost

The substrate left after the mushrooms have been harvested is known as spent compost. These substrate with innumerable mushroom threads (collectively referred to as mycelia) will have been biochemically modified by the mushroom enzymes into a simpler and more readily digestible This is present in large amounts, and raises the question of what can be done with it. It is certainly not desirable to leave it as a possible source of pollution. Mushroom mycelia can produce a group of complex extracellular enzymes which can degrade and utilize the lignocellulosic wastes in order to reduce pollution. It has been revealed recently that mushroom mycelia can play a significant role in the restoration of damaged environments. Saprotrophic, endophytic, mycorrhizal, and even parasitic fungi/mushrooms can be used in mycorestoration, which can be performed in four different ways: mycofiltration (using mycelia to filter water), mycoforestry (using mycelia to

Table 2.6: Nutraceutical Potential of the some Important Mushrooms

Mushroom Species	Active Constituents	Type of Polysaccharides	Medicinal Properties
Agaricus bisporus	Lectins	Heteropolysaccharides	Insulin secretary, anti-aging
Ganoderma lucidum	Polysaccharides, triterpenoids, germanium, nucleotides and nucleosides, ganoderic acid, beta-glucan	Heteropolysaccharides	Immunogenic, liver protective, antibiotic, cholesterol synthesis inhibitory; immunomodulatory, anti-cancerous
Lentinula edodes	Eritadenine, lentinan	Heteropolysaccharides	Cholesterol modulatory, anti-cancerous
Cordyceps sinensis	Cordycepin	Heteropolysaccharides	Lung infections curable, hypoglycemic, anti depressant
Auricularia auricula	Acidic polysaccharides	Homopolysaccharides	Anti-tumoureous, cho- lesterol triglycerides, and lipid levels modulatory, hypoglycemic, beneficial in coronary heart disease, immunogenic
Flammulina velutipes	Polysaccharide, flammulin, FVP (flammulina polysaccha- ride protein), peptide glycans, prolamin (active sugar protein), proflamin (glycoprotein)	Heteropolysaccharides	Antioxidant, anti-cancerous, anti-ageing; immuno- modulatory, anti-viral
Grifola frondosa	Grifloan, lectins	Heteropolysaccharides	Insulinogenic, hypoglycemic, ovulatory
P. florida		Homopolysaccharides	Anti-hyperglycaemic; anti- hypercholesterolemic
Pleurotus sajor-caju	Lovastatin polysaccharide	Homopolysaccharides	Cholesterol modulatory, cardio-vascular disorders preventive
Trametes versicolor	Polysaccharide-K (krestin), coriolon and glycoproteins	Heteropolysaccharides	Anti depressive, anticancerous, anti- Candida albicans, anti- viral, HIV replication inhibitory, hepatoprotective
Volvariella volvacea	Glycoproteins	Heteropolysaccharides	Cardio-protective, blood pressure reducing
Agaricus bisporus	Lectins	–	Enhance insulin secretion

restore forests), mycoremediation (using mycelia to eliminate toxic waste), and mycopesticides (using mycelia to control insect pests). It is known that there still remains in the spent compost a considerable amount of lignocellulosic material in addition to the mushroom mycelia and also other products formed by the metabolic activities of the mycelium. SMS is a good nutrient source for agricultural land and has high cation exchange capacity, capable to hold nutrients in the soil and to retain slow mineralization rate quality as an organic matter. The addition of SMS in the nutrient poor soil leads to improvement in soil texture, water holding capacity and nutrient status. SMS incorporation in soil also does not have any adverse effect on its alkalinity, rather, its amendment in soil leads to an increase in both pH as well as the organic carbon content. SMS contain 1.4 per cent nitrogen, 0.4 per cent phosphorus and 2.4 per cent potash. Thus, the spent compost should be capable of supporting further biological activities and used as following purposes;

☆ The growth of another species of edible mushroom

☆ Use as fodder for livestock.

☆ Used as a soil conditioner and fertiliser because it is a good nutrient source.

☆ Bioremediation because it is a good source of enzyme.

Chapter 3

General Morphology of Mushroom

Mushrooms can be defined as a macro-fungus with distinctive fruiting bodies, epigeous or hypogeous, large enough to be seen with naked eyes and picked up by the hands. The fruiting bodies of mushroom are umbrella shape like or of various other shapes, size and colour. Commonly it consists of a cap or pileus and a stalk or stipe but other variety have additional structures like veil or annulus, a cup or volva. Cap or pileus is the expanded portion of the carpophore (fruit body) which may be thick, fleshy, membranous or corky. On the underside of the pileus, gills are situated. These gills bear spores on their surface and exhibit a change in colour corresponding to that of the spores. The attachment of the gills to the stalk helps in the identification of the mushroom species. The edible fruiting bodies are called mushroom and poisonous ones are called as toadstool. Majority of these mushrooms belong to Hymenenomycetes of Basidiomycota. Fungus body made up by thallus which is undifferentiated of root, stem or leaf. Morphologically thallus of mushroom is divided into two parts:

- A. Underground part
- B. Aboveground part

(A) Underground Part

Underground part of thallus of mushroom is secondary mycelium which grows saprophytically. Secondary mycelium is called dikaryotic mycelium. Dikaryotic mycelium is therefore the result of the joining and melting of two sorts of monokaryotic mycelia. Only dikaryotic mycelium can form mushroom.

(B) Aboveground Part

The aboveground part of thallus of mushroom is fruiting body (tertiary mycelium): an edible part.

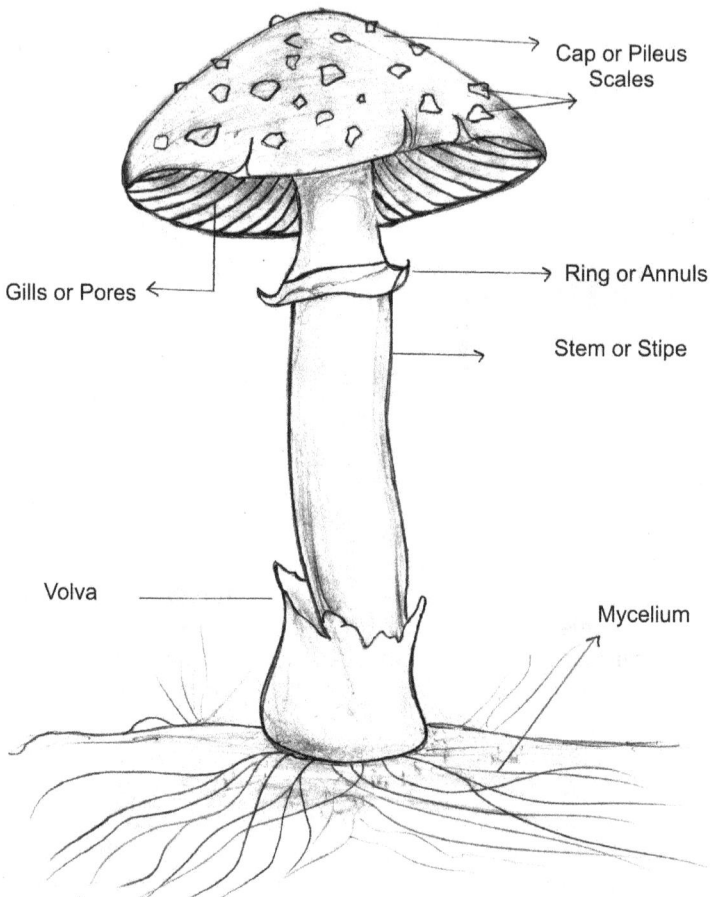

Figure 3.1: Different Parts of a Typical Mushroom.

The fruiting body/basidia consist of the following parts

(1) Pileus or Cap

It is various in shape and size depending on the species and the stage of growth. It is a convex, fleshy structure, which later becomes flat. The surface of the cap may be smooth, hairy or rough like fragments. The cap supports and protects the gills or pore where the spores are produced. The cap is an important character for identification of mushroom genera.

(2) Gills or Pores

These are tiny smooth, wrinkled or veined packed closely together forming a sponge layer usually present underside of the cap and produce spores. The gills are of different lengths and bear basidia all over the surface. Some mushrooms have pores instead of gills.

(3) Annuls/Ring

A partial veil grows from the edge of the cap to the stem. The ring is left on the stem as the cap grows and breaks the veil. The veil provides extra protection for the spores when the toadstool was young.

(4) Stipe/Stalk/Stem

The stalk or stipe is negatively geotropism and helps to hold up the cap or pileus, above the ground level to enable the vertically-falling spores to be easily drifted away by wind. The stem may be solid and fleshy.

(5) Volva

It is a cup like structure surrounding the swollen basal portion of the stalk of some mushroom. Volva is mostly found in poisonous mushroom such as Amanitas group *etc.* except some cultivated mushroom of *Volvariella volvacea.*

(6) Mycelium

The mycelium is the hidden 'body' of the fungus, which remains underground and grows saprophytically. It is the dikaryotic, secondary mycelium. It finds food for the fungus fruit.

Chapter 4

Life Cycle and Strain Improvement (Breeding Programme) of Mushroom

Life Cycle of Mushroom

If a section of the gills is cut and examined under the microscope, spores will be observed on their surface of gills. The spores will start to fall as the cap fully expands at full mature growth of mushroom fruit. These spores are very minute and fall to the ground usually with rain through the wind. Under the favourable conditions (optimum temperature and moisture), the spores will germinate to form a mass of mycelium. This is the vegetative phase of the mushroom fungus. The mycelium developing from the germinating spore is the so-called primary mycelium and is usually uninucleate and haploid. This stage is short-lived because mycelia from different spores tend to ramify and fuse to form the secondary mycelium with two compatible nuclei, which continues to grow vegetatively and is able to form fruiting bodies (Figure 4.1). Secondary mycelium is septate and, since each cell contains all the necessary organelles for independent growth, fragments of the mycelium can generate to form new colonies.

Through the mycelium, mushroom fruit absorb food from the substrate on which they grow. The mycelium produces enzymes that digest complex carbohydrates, lipid and protein, which are then easily absorbed by the hyphae.

Mushroom completes its life cycle through two phases, vegetative and reproductive. Vegetative growth indicates linear growth of fungal mycelia dissolving complex substrate components into simpler molecules and absorbing them as nutrients. It cease vegetative growth under low temperature, high humidity, much oxygen and light. The cease in vegetative growth results in beginning of fruiting

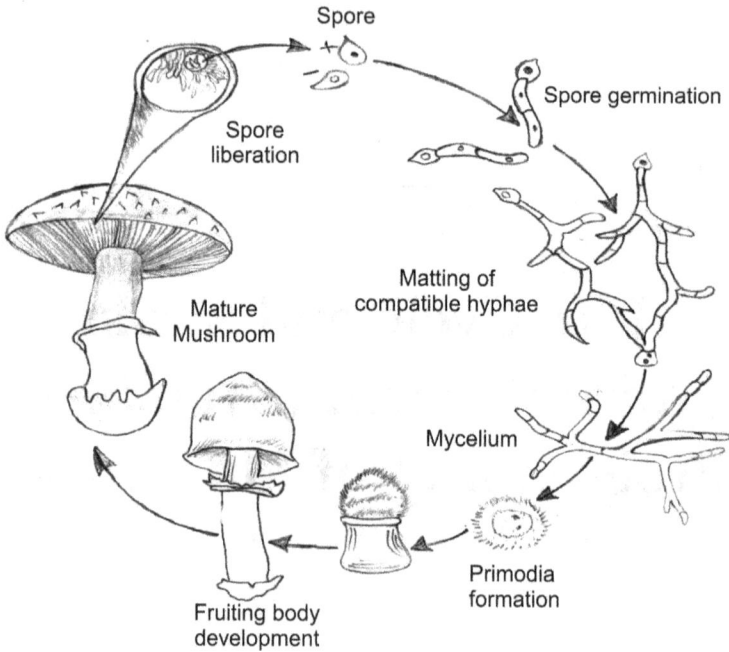

Figure 4.1: Life Cycle of Mushroom.

bodies and this is called reproductive growth. Thus mushroom cultivation can be called as practice of obtaining fruit bodies through modulating these both stages.

Strain Improvement (Breeding Programme) of Mushroom

Most edible cultivated mushrooms belong to the Agaricales group of Basidiomycetes. The mode of sexual reproduction involves plasmogamy, karyogamy and meiosis which result in the formation of four haploid spores. The spores resulting from sexual reproduction are called basidiospores. The reservoir of edible mushrooms, like other micro-organisms for industrial purposes, is not unlimited. It is generally recognized that in order to maintain and breed high-yielding strains, the techniques employed in mushroom breeding should now and then be modified and improved in accordance with new scientific progress and findings, as a whole, and in microbiology and genetics in particular.

1. By Selection

Selection can be made from multi-spore or single-spore cultures. Tissue culture of selected sporophores has also been used to fix desired variants. In the short term, selection alone appears to have some role to play in mushroom strain improvement, but genetic improvement by selection becomes increasingly difficult. Therefore, strain improvement through hybridization has become a recourse and also necessity.

2. By Hybridization

Hybridization seems to offer the best improvement breeding technique, especially with regard to multiple gene transfers mediated by protoplast fusion. In addition to the conventional method of matings between two genetically compatible strains through which dikaryon mycelia and fruiting bodies are formed, steps towards a broader spectrum of hybridization can be achieved in strain improvement of edible mushrooms by the following ways:

A. Use of Auxotrophs

Auxotroph can be obtained naturally or induced by mutagenesis. The contrasting auxotrophs can be paired and the products can be screened for hybridity on minimal medium. Certainly, the feasibility of auxotrophs to be used as a tool for hybridisation depends on how easily auxotrophs can be obtained in the strains of the mushroom.

B. Use of Resistance Markers

Mutants resistant to anti-metabolites have been suggested recently as alternatives to auxotrophs for use in mushroom breeding programmes. The treated spores or hyphal fragments, which can grow on a medium containing an inhibitory concentration of the anti-metabolite, would be considered to possess the marker. Complementary resistant strains would be grown together and, hybridity can be confirmed by transferring it to a medium containing the two appropriate anti-metabolites.

C. Protoplast Fusion

One of the most effective barriers to sexual reproduction is the inability of hyphae from two selected strains to fuse. Several laboratories have reported that protoplasts can be isolated from plant and microbial cells by enzymatic breaking of the cell wall, in the presence of an osmotic stabilizer. Such protoplasts can be effectively induced to fuse in the presence of polyethylene glycol (PEG). After a short period of time, protoplasts can regenerate their cell walls and start to propagate as normal cells or hyphae. These cells are heterokaryotic if fusion occurred between cells from genetically different strains. This can serve to increase the frequency of intra-species crosses in organisms in which natural matings rarely occur. The technique has even a much wider application, and can be used for inter-species and inter-generic crosses in some organisms, which normally cannot be crossed. Although such an approach has been carried out in several laboratories, until now, no clear and economically applicable results have been reported in edible mushrooms.

Specific Technique for Development of New Lines

1. Selection from Monospore or Single Spore

This technique is better for obtaining new lines because it varies in rate of growth, shape of fruiting bodies and productivity. It is used to develop new lines only for *Agaricus bisporus* and *Volvariella volvacea*.

Preparation of Single-Spore Culture

A mushroom fruit with the cap is cleaned and laid flat on a filter paper in a sterilized Petri plates. This is then covered with the help of a beaker and incubated in BOD at 25±1°C for 1to 2 days. The veil opens and discharges the spores, onto the filter paper. The beaker is removed and a sterilize lid is placed on the dish. The spore print is stored in the refrigerator for future use. In case of *Volvariella* and *Pleurotus*, the fruits used to obtain the spores should be newly opened. Spore print should be stored at room temperature because low temperature reduces germination.

When ready for use, the filter paper with the spore print is aseptically cut into strips and one small strip is placed in 10 ml of sterile distilled water. For heavy spores' loads, a dilution series may be made. A suspension (0.1 ml) of the spores is then planted on a Nutrient medium in Petri dishes and incubated at desired temperature. Spores will start to germinate 5 to 7 days later and can be observed by examining the Petri dish under microscopes. After making with a marker pen, the germinated spores are transferred, with a portion of the agar, to a new medium in a test tube or a Petri dish

2. Selection from Multi-Spore Cultures

A spore suspension, collected as described above is mixed with warm and liquefied agar medium. The agar is then allowed to solidify. Fused mycelium will begin to grow on the surface of the agar after and during 3 to 5 days. Another method is dipping a sterilized needle in sterile water, then touching the moistened tip to a spore suspension or print. The mass of spores is then streaked directly on a solidified agar surface and resulting fused mycelia are allowed to grow on the surface. A piece of the mycelial plug is transferred to new agar medium and the first selection is made after one week.

3. Selection from Simple Mixing

Two fertile strains are obtained through single-spore or multi-spore culture. These are inoculated close to each other on an agar plate and grow together. The mycelium, along the lines of juncture, is inoculated in spawn bottles to prepare spawn for future use.

Chapter 5

Identification of Mushroom

Proper identification requires the basic knowledge of the structure of the fungus and its characteristics. Identification of given mushroom requires examination of fruiting bodies carefully. A fresh fruiting body is better than preserved or a dried one. Identification of mushroom fungus is based on the following traits.

☆ Size, shape, texture and colour of the cap.

☆ Size, shape, texture, colour and consistency of the stalk.

Smooth

Velvety

Hairy or fibrous

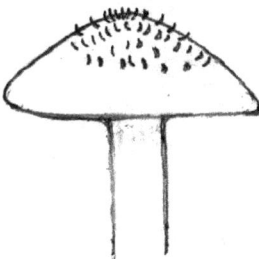
Raised scales

Flat scales

Patches

Figure 5.1: Identification of Mushroom Species by Upper Surface of the Cap.

☆ Presence of gills, pores or spikes under the cap.

☆ Mode of attachment of the gills to the stalk.

☆ Gills morphology.

☆ Spore colour in mass.

☆ Mushroom change colour when cut or bruis

☆ Chemical test or reaction: It is desirable to go for "spore print" examination to determine the real colour of the spore prints can be made.

Spore Print

A spore print is a piece (tinfoil or glass slide) used to collect the spores of a particular mushroom strain. To start growing mushrooms one must first get a spore print. A spore print can be obtained from a mushroom found in nature.

The stipe of the mushroom fungus is cut close to the cap and the cap is placed it gills down on a piece of clean white paper. Black paper is in case of species where spore are white or light colored. If a specimen is partially dried, a drop or two of water is added to the cap surface to help the release of spores. The cap is covered with an inverted tumbler or dish to exclude air and prevent the cap from drying out. Spores begin to eject from the gills during 2nd to 24th hours. The cap is then carefully lifted from the paper. Spores of the mushroom get deposited in a thin layer on the paper making a perfect pattern showing the arrangement of the gills (Figure 5.2). Using, gummed paper, one can obtain fairly permanent spore print and preserve for future references. The color of spore print vary from species to species and this is dark brown in case of *Agaricus,* white to lilac-gray in case of *Pleurotus* and pink in case of *Volvariella volvacea.*

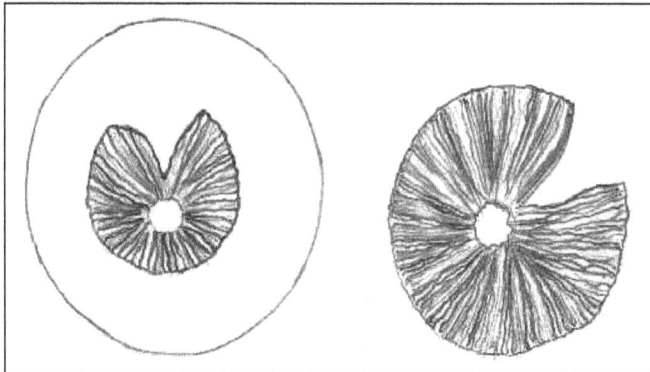

Figure 5.2: Identification of Mushroom Species by Spore Print Method.

Mode of Attachment of the Gills

The mode of attachment of the gill to the stalk indicates the genus of the mushroom. The mushroom is cut longitudinally through the cap, exposing the point of attachment of the gills to the stalk in order to determine the mode of attachment. The different patterns by which the gills are attached to the stem are shown in

(Figure 5.3). The environment in which the mushroom is growing should be noted. It is important to know whether the mushroom grows directly on the ground, on decaying wood, on living tree trunk or on compost. The species of tree on which the mushroom are found growing or the type of grasses or moss present in the area where the mushroom are collected should also be noted.

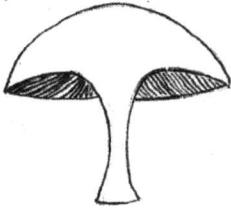

Adnate
(Wide attachment of
gills with stem)

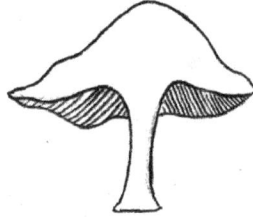

Sinuate
(Gills smoothly notched and
running down stem)

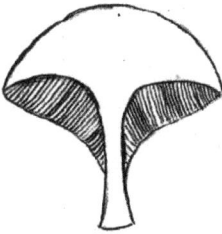

Decurrent
(Attachment of gills along the
stem for up to 2-4 cm)

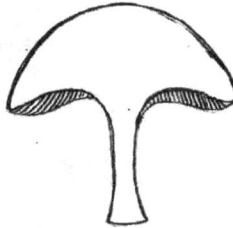

Adnexed
(Narrow attachment of
gills with stem)

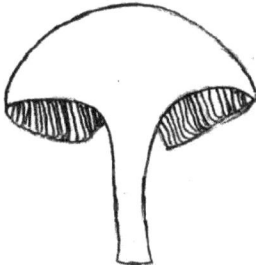

Emarginate
(Notched gills just before
attachment to stem)

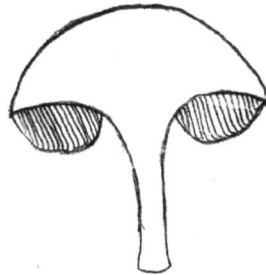

Free
(Separated gills from stem)

Figure 5.3: Identification of Mushroom by Gill Attachments to the Stock.

Table 5.1: Identification of Mushroom Fungus Based on Morphological Characteristics

Mushroom Species	Morphological Characteristics for Identification
Agaricus	Agaricus are characterized by having a fleshy cap or pileus, which are convex to broadly convex or nearly flat in age, dry, smooth or with pressed-down fibers or small scales, white in some varieties, brown in others. Gills are free from the stem, close, pinkish to pinkish brown at first and later becomes dark brown to blackish. Stem or stipe are 2-8 cm long, 1-3 cm. thick, smooth or with small scales.
Pleurotus	Pleurotus are characterized by having broad, fan or oyster-shaped cap with different color according to which is laterally attached (with no stem). The flesh is white, firm, and varies in thickness due to stipe arrangement. The gills of the mushroom are white to cream, and descend on the stalk if present. If so, the stipe is off-center with a lateral attachment to wood. The mushroom's stipe is often absent. When present, it is short and thick.
Volvariella	Volvariella are characterized by having deep salmon pink gills. They lack a ring, and have an Amanita like volva at the stem base.
Calocybe indica	It is robust, fleshy, milky white with a long stipe and a small pileus, umbrella like, resembling button mushroom.
Lentinula edodus	Shiitake are tan to dark brown and have broad umbrella shaped cap, wide open veils, tan gills and curved stipe.
Ganoderma lucidum	This is kidney-shaped with pores instead of gills on its underside. The upper surface is shiny and dark red. When it is young, it has yellow, white edged and relatively smooth upper surface. As it matures, the entire upper surface gets reddish brown and more scalloped.
Boletus	This is easy to identify as they don't have gills but 'spongy' pores which are white, cream or yellow.
Amanitas	Mushroom of this group has white gills and spores and mostly grow from a sac like or bulbous structure called a volva.
Russulas	They have very brittle gills and stems.

Chapter 6

Preparation of Culture Media, Preservation and Storage of Mushroom Cultures

Culture Media

Mushrooms grow on many culture media with different agar formulations (both natural and synthetic), depending on the variety of mushroom to be cultivated and its purpose. Following culture media are recommended for pure culture, isolation and enumeration of mushroom fungus.

A. Preparation of Potato Dextrose Agar (PDA) Medium

The ingredients/litre of deionised water

Peeled and sliced potato	:	200.0 g
Dextrose	:	20.0 g
Agar-agar	:	20.0 g

Method of preparation

☆ Required amount of peeled potato is cut into fine pieces.

☆ And it is boiled in 500 ml of distilled water for 30 minutes, thereafter potato extract is filtered through muslin cloth.

☆ 20 g of dextrose and 20 g of agar-agar is dissolved in 500 ml boiling water.

☆ Potato extract is added in boiling mixture to make 1000 ml and mixed thoroughly by stirring with glass rod.

☆ After few minutes of boiling it is transferred to about 200 ml in each 500 ml capacity flask (2/3 of its capacity) and the pH of the medium is adjusted 7.0 (±0.2) by adding NaOH or acetic acid.

☆ Flasks are plugged with non-absorbent cotton and covered with aluminum foil before autoclaving.

☆ Medium filled flask is autoclaved under 15 pounds pressure per square inch (p.s.i.) at 121.6°C for 15 minutes.

☆ Agar medium is allowed to cool and solidify before use.

B. Preparation of Malt Extract Agar (MEA) Medium (Recommended by Thom and Church, 1926)

The ingredients per litre of deionised water

Malt extract	:	200.0 g
Dextrose	:	20.0 g
Peptone	:	6.0 g
Agar	:	20.0 g

Method of Preparation

☆ Required ingredients are added in 1000 ml of distilled water as per quantity given above.

☆ Agar is added and allowed to boil for 10-15 minutes and mixed thoroughly by stirring with glass rod. Thereafter it melts down and allowed to be sterilized.

☆ 1 per cent peptone or 0.5 per cent yeast may be added for faster mycelial growth in both PDA and MEA media.

C. Lambert's Agar Medium

The ingredients per litre of deionised water

Glucose	:	10.0 g
Mangnesium sulphate	:	0.5 g
Potassium dihydrogen	:	1.9 g
Agar-agar	:	20.0 g

Method of Preparation

☆ Required amount of ingredients are added in 1000 ml of distilled water as per quantity given above.

☆ After few minutes of boiling, 200 ml of the medium is transferred in each flask of 500 ml capacity (2/3 of its capacity). pH of the medium is adjusted 7.0 (±0.2) by adding NaOH or acetic acid.

☆ Flasks are plugged with non-absorbent cotton and covered with aluminum foil, autoclaving at 15 pounds pressure per square inch (p.s.i.) at 121.6°C for 15 minutes. Thereafter, agar medium is allowed to cool and be solidified before use.

D. Wheat Extract Agar Medium

The ingredients per litre of deionised water

Wheat grain	:	10.0 g
Agar-agar	:	20.0 g

Method of Preparation

☆ Thirty two grams of wheat grains is added in one litre of distilled water for 2 hours and is filtered after 24 hours. Thereafter, 20 g agar is added to one litre of filtrate and allowed to boil.

Figure 6.1: Culture Media for Pure Culture of Mushroom Fungus.

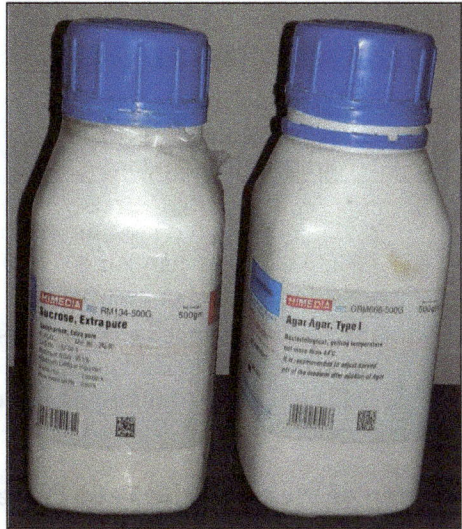

Figure 6.2: Commercial Sucrose and Agar-Agar.

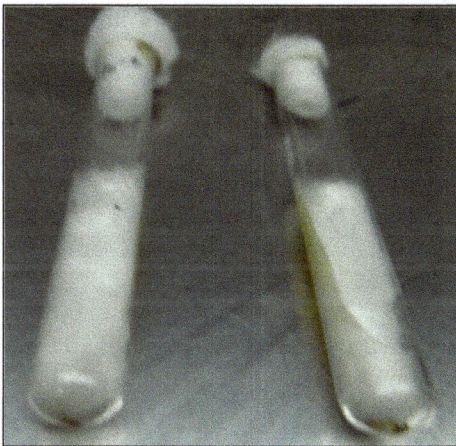

Figure 6.3: Mushroom grown under Slant Culture for Storage.

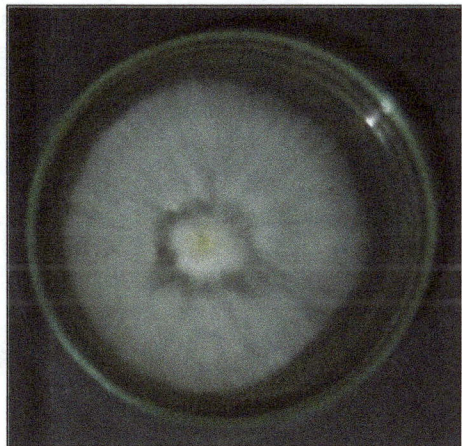

Figure 6.4: Mushroom Mycelium in Culture Plate.

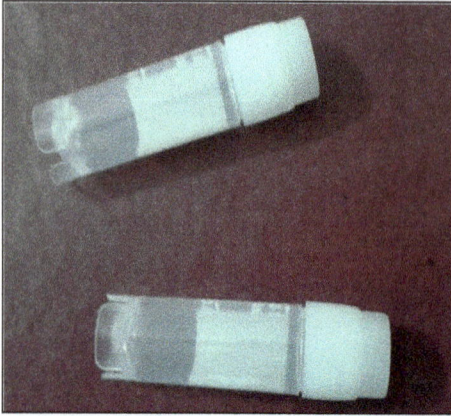

Figure 6.5: Plastic Ampules Used to Preserve Mycelium in a Liquid Nitrogen Refrigerator.

Figure 6.6: Grain Substrate in Bottles for Storage of Mycelium.

☆ 200 ml prepared medium are transferred in each flask of 500 ml capacity (2/3 of its capacity) and the pH of the medium is adjusted 7.0 (±0.2) by adding NaOH or acetic acid.

☆ Flasks are plugged with non-absorbent cotton and covered with aluminum foil before autoclaving under 15 pounds pressure per square inch (p.s.i.) at 121.6°C for 15 minutes. Agar medium is allowed to cool and get solidified before use.

Plating of Media in the Petri Plates and Slanting

About 15 ml of molten PDA or MEA media is transferred in Petri plate. Water vapur is often get condensed forms on the inside surface of the upper lid of a Petri dish under the agar medium gets cool down. To reduce condensation, one can wait a period of time before pouring. If culture tube is used 10 ml of the liquid agar solution is added and slanted to the tube.

Culturing the Mushroom

Mushroom culture may be initiated in two ways, one from its spores and the other from living tissue of a mushroom. Either type can produce a viable strain of mycelia. Mostly spores are used to develop mushroom culture. Spores have the advantage of being viable for week even after mushroom gets decomposed.

1. Tissue Culture

A well grown mushroom with membrane covering the gills is selected and from which a small bit of membrane from gill portion is taken by the help of a sterilized needle and inoculated centrally on the surface of the pre sterilized PDA or MEA culture medium. The Petri-plates are incubated at 25±2°C in BOD incubator for one week. Mycelium from growing edges is carefully transferred to MEA/PDA slants and again incubated for 2-3 weeks to obtain pure cultures.

2. Spore Culture

Collect aseptically large sized healthy mushroom with membrane (veil) still intact, surface sterilize the mushroom, mount on a wire stand over a Petri dish under glass beaker already sterilized in an oven at 160°C for about 1 hour and cooled. The spores get deposited as spore print. The spores are stored under sterile condition in a refrigerator for future use. The Spores are inoculated to the sterilized PDA and MEA culture slants. These slants are then incubated at 25 ± 2°C for 2 weeks to obtain pure culture.

3. Multispore Culture

For raising culture, spore suspension is prepared in sterilized distilled water. One ml of spore suspension containing more than hundred spores is mixed in each culture tube containing about 5-7 ml of sterilized in PDA and slants are prepared. The slants are incubated at 28°C. The mycelium threads become visible on slant surface.

4. Cultures from another Source

Cultures also may be grown from spawn obtained from another source. A piece of the spawn is aseptically transferred to agar slants. However, this is risky because the number of transfers that the spawn culture has undergone is rarely known. If this procedure is followed, it may be advisable to first grow the spawn into fruiting bodies, and then make the necessary isolations from the fruiting body.

Preservation and Storage of Cultures

Culture preservation is indispensible to have a stock of mycelium and for smooth spawn operation. The main objective of culture preservation remains to store cultures in a viable and stable form for longer periods without losing genotypic, phenotypic and physiological traits. Most mushroom cultures are maintained on a suitable agar medium in glass tests tube. The tubes are sealed with melted paraffin wax or wrapping the cotton wool plug with parafilm which allows for satisfactory gas exchange.

The most common method of short-term storage of mushroom culture is the storage of culture tubes either at room temperature (25–30°C) for a period of 1–2 months or in refrigerator (5–8°C) for an average period of 3–4 months. More ordinary refrigeration of cultures at -10°C or lower if possible, will allow storage of mycelium more than year with little loss in viability.

Freezing methods of culture preservation are most effective methods are freeze-drying (lypholization) and freezing and storage in liquid nitrogen.

1. Free-Drying and Freezing

This is most economical and effective method under which the spores of mushroom fungi remain dormant and viable for longer period. The spores are frozen and at the same time dried under low pressure in vacuum to remove water from spores. For freeze drying, strains of mushroom grown on Petri plates containing suitable agar medium. Three plugs of the advancing age of the culture are removed with the help of sterile cork borer and transferred in heavy walled borosillicate glass ampules for freeze-drying and storing in liquid nitrogen. The disadvantage of this method is to cause freeing injury to biological system of mushroom fungus.

2. Cryogenic Freezing

A cryoprotectant is a substance that is used to protect biological tissue from damage during freezing. Common cryoprotectants include glycerol and dimethyl sulfoxide (DMSO). Both of these are penetrating cryoprotectants at physiological temperatures, meaning that they are able to move across cell membranes into the cytoplasm. Cryoprotectants serve to help maintain osmotic equilibrium during the freezing process and decrease the damage caused by intracellular ice nucleation. Glycerol is commonly used to protect mushroom cultures from freeze damage. We commonly add a 1 ml sterile glycerol solution (10 per cent) to 2 ml capacity cryotubes and then submerge a few (up to 6) mycelial plugs or spawn grains into the solution.

Liquid nitrogen refrigeration is now considered the only adequate method to preserve mushroom cultures. The viability of samples frozen in liquid nitrogen can vary depending on: i species of mushroom, ii. Mycelium age of fungus, iii. growth stage, iv. type of cryoprotectant used, iv. rate of cryoprotectant penetration, v. method and rate of freezing, and vi. thawing time and temperature.

Cultures stored in mineral oil can be kept either at room temperature or in a refrigerator for 1 to 3 years. Cultures of *Volvariella* do not survive in mineral oil, or at temperature below 10°C. Hence, they are kept only in incubator at 15-20°C. The only long-storage method of *Volvariella* cultures is liquid nitrogen storage.

Chapter 7

Techniques of Pure Culture of Mushroom

Pure Culture of Mushroom

The pure mushroom culture can be obtained from either from spore or tissue. The advantage of using spores is that they are viable for weeks to months after the mushroom has decomposed. If the specimens are several days old, too dry or too mature, a pure culture will be difficult to isolate.

A. Pure Culture from Spore (Spore Print)

- ✰ Once a spore print is obtained, mushroom culture can begin.
- ✰ Sterilize an inoculating needle by holding it over the flame of an alcohol lamp for 5 or 10 seconds until it is red hot.
- ✰ Cool the tip of needle by inserting it into the sterile media in a Petri dish and scrape some spores off the print.
- ✰ Transfer the spores by streaking the tip across the agar surface.
- ✰ Incubate the Petri dish at desired temperature.
- ✰ When starting a new culture from spores, it is best to inoculate at least three media dishes to improve the chances of getting a successful germination.

Once the mushroom mycelium has been identified, sites of germinating spores should be transferred to new media dishes. Continue transferring the mycelium away from the contaminants until a pure strain is established.

B. Pure Culture from Tissues

Tissue from mushrooms fruiting body must be taken within twenty-four to forty-eight hours of being picked. Since the entire mushroom is composed of compressed mycelia, a viable culture can be obtained from any part of the mushroom fruit body. The cap, the upper region of the stalk and/or the area where the gill plate joints the underside of the cap are the best locations for excising clean tissue.

☆ This skin can be peeled back and a tissue culture can be taken from the flesh underlying it.

☆ Wipe the surface of the mushroom with a cotton swab soaked in alcohol and remove any dirt or damaged external tissue.

☆ Break the mushroom cap or stem, exposing the interior hyphae.

☆ Immediately flame a scalpel until red-hot and cool in a media filled Petri dish.

☆ Now cut into the flesh removing a small fragment of tissue.

☆ Transfer the tissue fragment to the canter of the PDA filled Petri dish as quickly as possible, exposing the tissue and agar to the open air for a minimal time.

☆ Repeat this technique into at least three, preferably five more dishes.

☆ Label each dish with the species, date, type of culture (tissue) and kind of agar medium.

☆ Incubate the Petri dish at desired temperature.

If successful, mycelial growth will be evident in 3 to 7 days. An overall contamination rate of a 10 per cent is one most cultivators can tolerate. The contaminants often appear near to the point of transfer. Their numbers depend on the cleanliness of the tissue or spores transferred and the hygienic state of the laboratory where the transfers were conducted.

Sectoring and Abnormalities in Cultures

Any type of mycelial growth that differs in appearance, growth rate and colour pattern from the typical appearance of a given strain is called sectoring. It is often seen as a rapidly growing area near the leading edge of growth in an agar slant, which often exhibit a different growth habit than the rest of the culture. A sector may or may not revert to normal growth. Other abnormalities that might appear in a culture are fluffy, aerial mycelium or a colour change such as browning or darkening of the mycelium. The causes of sectoring are not well understood yet. But it is believed that sectoring represents some type of genetic change, perhaps chromosomal aberration, somatic recombination, or chromosomal loss in the culture. Presently, no method is available to determine the ultimate productivity of a culture on agar. Cultures are propagated solely on the basis of cultural characteristics exhibited in an agar slant. These characteristics vary considerably from strain to strain, but it has been noticed that transfers from typical, uniform cultures of a given strain will produce reliable spawn when handled properly. On the other hand, a sector or change in vegetative growth could affect the productivity of the culture to an unknown extent. Therefore, it is very important to recognize and avoid propagation of sectored mycelium.

Table 7.1: Problems, Causes and Solution during Pure Culture

Sl.No.	Problem	Causes	Solution
1.	Agar medium very soft or scarcely solidifies.	Quantity of agar insufficient *i.e.* too low.	Use proper quantity of agar in medium.
2.	Agar surface in the plates not smooth or lumpy.	Agar medium partially solidify when poured.	Pour agar medium when it is still hot.
3.	Contaminants appear after 2-3 days on the surface of the medium after sterilization and before inoculation.	Medium not sufficiently sterilized. Medium not aseptically poured.	Sterilization should be carried for the recommended period and temperature/pressure. Medium should be poured aseptically.
4.	Transferred mycelial bit/tissue resume no growth.	Non viable inoculumn/culture. Wrong type of medium. Incorrect formulation or pH. Needle or scalpel used to transfer the culture bit too hot.	Use viable culture/actively growing culture. Use correct medium. Properly check the formulation and pH of the medium. Cool the flamed needle before picking the inoculums.
5.	Contamination develops on the plugs after 2-3 days.	Culture used already contaminated. Filters of the laminar flow damaged. Incubation too much loaded with air borne inoculum.	Use fresh/disease free cultures. Filters should be checked or replaced as per recommendation. Sterilize the incubation rooms from time to time.
6.	Resulting mycelial growth slow and fluffy or feathery.	Strain degenerated.	Obtain another culture or retrieve stock culture.

Care and Handling of Cultures

Following points must be taken under consideration before starting culture to minimize the chances of unwanted problems during spawn production process.

☆ Opening of mailing containers used to ship agar cultures immediately to allow maximum air circulation and gas exchange after receipt.

☆ Avoiding storage of culture in closed containers since volatile substances produced by the fungus can inhibit mycelial growth.

☆ Distribution of uniform and normal mushroom cultures among growers, which should have similar growth rates to other isolates of the same strain.

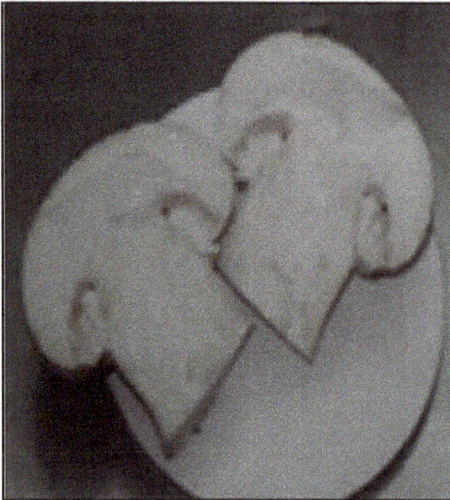

Figure 7.1: Tissues a Source for Pure Culture.

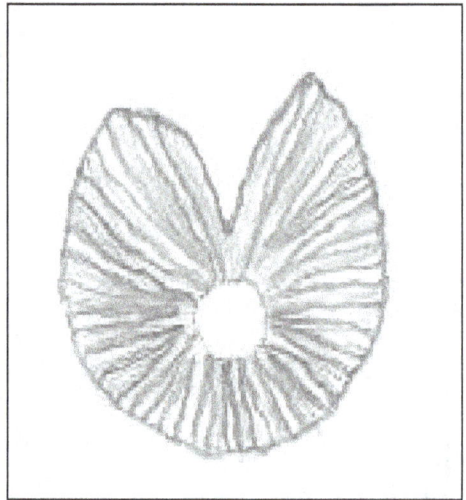

Figure 7.2: Spore Printing for Pure Culture.

Figure 7.3: Preparation of Culture Plate.

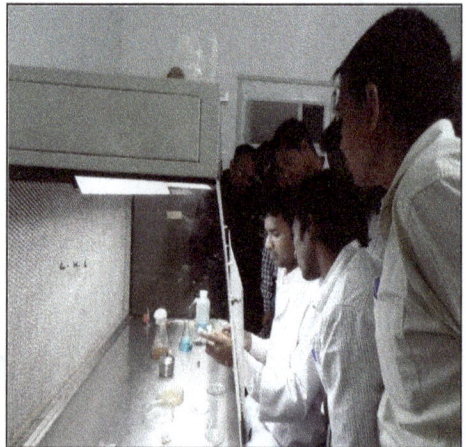

Figure 7.4: Inoculation of Mushroom in Petri Plate.

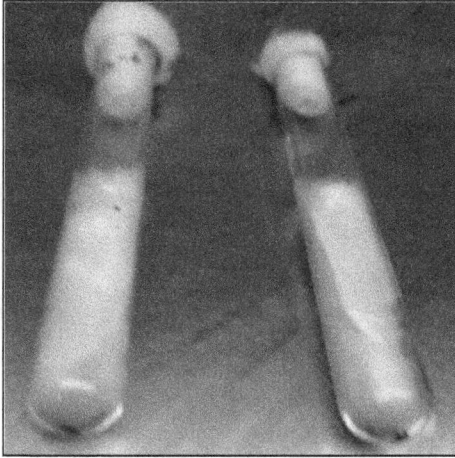

Figure 7.5: Test Tube Containing Pure Culture of Mushroom.

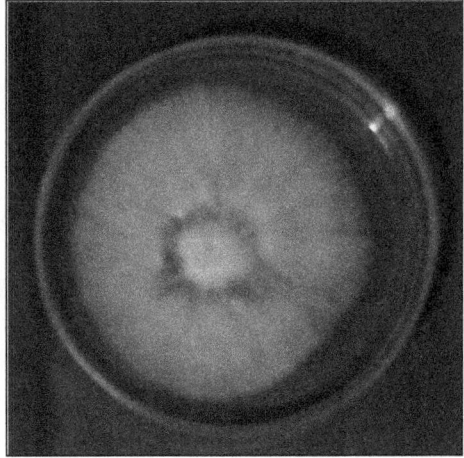

Figure 7.6: Petri Plate Containing Pure Culture of Mushroom.

☆ Use of commonly used agar medium (potato dextrose agar, potato dextrose yeast extract agar and malt extract agar) for maintaining pure cultures of the cultivated mushroom.

☆ Transfers into fresh agar medium as soon as possible, usually within 1 to 2 weeks. However, agar cultures should be grown for 2 to 4 weeks to determine whether typical characteristics remain stable before use.

☆ Incubation of new culture at about 23°C for maximum growth.

Chapter 8

Techniques of Spawn Production

Spawn

A steam sterilized grain colonized by the mycelium of the mushroom and used to "seed of mushroom. Spawn is the vegetative mycelium from a selected mushroom grown on a convenient medium like wheat, pearl millet, sorghum, rye etc for raising mushroom crop or mushroom spawn is the mushroom mycelium growing on a given substrate. The yield and quality of spawn is governed by the genetic makeup in the strain and the technology including the substrates used in spawn production.

Kinds of Spawn

Spawn which are used for mushroom cultivation are given below.

1. **Natural or virgin spawn:** Spawn from the meadows and pastures where mushroom naturally grow.
2. **Manure spawn:** Spawn on the compost substrate where mushroom grow.
3. **Horse spawn:** Spawn from the horse manure where mushroom grow.
4. **Flake spawn:** Spawn made by breaking down beds (6 to 8 inches diameter) through which the mycelium has run, before the mushroom crop appears.
5. **Tobacco spawn:** Tobacco stem as the basal medium for spawn making or a mixture of stem of tobacco with humus and peat adjusted to pH 6.0-6.7.
6. **Powder spawn:** Spawn made up of 20 different substrate, which was developed by Stoller in 1971.
7. **Grain spawn:** Grain spawn was first time introduced by Sinden in 1932 who added calcium salts to hard rye grain. Grain of wheat, rye, sorghum, bajra, rice, maize etc are used as a substrates for spawn production.
8. **Granular spawn:** Granular spawn was first time introduced by Hu and Lin in 1972 who used shell powder, compost powder, grain hull powder and starch.

Various type of spawn are being used for seeding the compost throughout the world, but now a day, grain spawn is preferred to rest of others. However, rye grains are preferred over the other grains in most of the foreign country, but wheat, sorghum and bajra are commonly used in India for spawn production.

The techniques of production of mother and commercial spawn for every species of mushroom are more or less similar. The most economic and farmers friendly technique is being described here.

Mother Spawn Production

Mother spawn prepared using pure culture mycelium is placed onto steam-sterilized grain, and in time the mycelium completely grows through the grain. This grain/mycelium mixture is called mother spawn. It is also known as stock or master culture.

Requirements

The required materials are listed below for mother and commercial spawns production

Sl.No.	Raw Materials	Sl.No.	Appliances/Tools
1.	Fresh mushroom culture	1.	Autoclave
2.	Grain substrate (wheat, jowar, bajra, rice)	2.	Laminar flow
3.	Poly propylene bag (PP bag)/Glucose/ milk bottle	3.	BOD Incubator
4.	Aluminium foil	4.	Inoculation needle/Spatula
5.	Chemicals ($CaCO_3$ and $CaSO_4$)	5.	Spirit lamp
6.	Non-absorbent cotton	6.	Thermometer
7.	Rubber bands	7.	Hygrometer and Humidifier
		8.	Sprit (70 per cent Alcohol)
		9.	Steel racks
		10.	LPG/electric stove (for boiling substrates)
		11.	Fry pan (capacity 20-25 kg) for boiling substrate
		12.	Mesh (for filter water from substrate-3x6 fit

Procedure for Preparation of Mother Spawn

1 Pure culture of fleshy mushrooms can be prepared either by multi-spore culture or tissue culture.

2 Healthy and cleaned cereal grains (wheat/jowar/bajra) may be used as substrate.

3 Grains are boiled in water (15-20 minutes). Normally for soaking and boiling 10 kg of wheat grain requires 20 litre water.

4. Excess water is drained off by spreading on sieve.

5. Wheat grains are dried on polythene or bed sheet in shade (4 hours) to remove extra moisture.

6. $CaCO_3$ (0.5 per cent) and $CaSO_4$ (2 per cent) are added and mixed in dried grain.

7. Glucose/milk bottle up to 3/4 volume is filled with 500g dry grains.

8. Bottle is plugged with non-absorbent cotton and is covered with aluminium foil before autoclaving at 126°C at 20 psi for 1.30 to 2 hrs.

9. Sterilized bottles are taken out from the autoclave while still hot and are shaken to avoid clump formation.

10 Sterilized bottles are immediately left in the inoculating chamber and allowed to cool down overnight.

11. Bottles are kept in laminar air flow chamber under U.V. tube for 20-30 minutes before inoculation.

12. Bottles are inoculated with two bits of agar medium colonized with the mycelium of pure culture of mushroom raised either by tissues or spores.

13. Inoculated bottles are incubated in BOD at 25±1°C for 20-22 days.

14. Inoculated bottles are gently shaken on 5th and 10th day.

15. Now, mother spawn is ready.

17. Mother spawn bottles must be marked with firm name, species, quantity, date of inoculation to know the age and type of spawn.

Precautions to be taken during Mother Spawn Production

☆ Always use fresh mushroom fungal strain.

☆ Inspect the bottles regularly and discard contaminated one immediately.

Figure 8.1: Mother Spawn.

☆ Observe that within 18-20 days of inoculation, mycelium growth covers entire substrate and the spawn are ready for use.

Commercial/Planting Spawns Production

Commercial spawn can be prepared in heat resistant polypropylene bags (pp bag) instead of glucose bottle as used in mother spawn production. Commercial spawn is prepared from mother spawn or master culture.

Requirements

The required materials used in commercial spawn production are same to that in mother spawns production except pp bag in place of glucose bottle.

Procedure for Commercial Spawn Production

1. Healthy and cleaned cereal grains (wheat/jowar/bajra) may be used as substrate.
2. Grains are boiled in water (15-20 minutes) (Normally for soaking and boiling 10 kg of wheat grain requires 20 litre water).
3. Excess water is drained off by spreading on sieve.
4. Wheat grains are dried on polythene or bed sheet in shade (4 hours) to remove extra moisture.
5. $CaCO_3$ (0.5 per cent) and $CaSO_4$ (0.2 per cent) are added and mixed in dried grains.
6. Heat resistance polypropylene bags are filled with 500g or 1 kg dry grains. For half and one kg spawn the bags should be of 35 x 17.5cm and 40 x 20cm size respectively.
7. PP bags are plugged with non-absorbent cotton with the help of a PP neck plastic ring and are covered with aluminium foil before autoclaving at 126°C at 20 psi for 1.30 to 2 hrs.
8. Sterilized PP bags are shaken well before inoculation so that the water droplets are accumulated inside the bags could be absorbed by the grains.
9. Sterilized PP bags are left in the inoculated chamber and allowed to cool down overnight.
10. Next day, bags are kept on laminar air flow chamber under U.V. tube for 20-30 minutes before inoculation.
11. 10-15 gm of grains from mother spawn bottle is inoculated into bags. Normally one bottle of mother spawn which is sufficient for inoculating 25 to 30 commercial spawn bags.
12. Inoculated bags are again shaken so that the inoculum is well mixed with other grains.
13. Inoculated bags are kept in incubation room at 25±1°C for 15-16 days for the spread of mycelia.
14. Inoculated bags are gently shaken on 5th and 10th day for proper colonization of fungal mycelium among the substrate.

Figure 8.2: Commercial/Planting Spawn.

Figure 8.3: Commercial/Planting Spawn Kept on the Shelves.

15. Commercial spawn is ready in 18-20 days.
16. PP bag must be marked with firm name, species, quantity, date of inoculation to know the age and type of spawn.

Precautions to be taken during Commercial Spawn Production

☆ Select high yielding, early producing and high spore farming strain.

☆ Always use fresh, good quality and unbroken seed grain (grain substrate).

☆ Always keep the inoculation chamber and its surrounding very neat and clean.

☆ Switch on UV tube in the inoculation chamber for 30 minutes before inoculation by keeping sterilizing substrate, forceps and culture inside the chamber.

Table 8.1: Problems, Causes and Solution during Spawn Preparation

Sl.No.	Problem	Causes	Solution
1.	Grains contaminated after sterilization and before inoculation.	Highly infected seeds. Grains not fully sterilized.	Use fresh and clean seed. Prolong sterilization period.
2.	Mycelial growth very thin and hardly penetrates the grains.	Grains too dry.	Boil the grains sufficiently. Adjust proper moisture levels.
3.	Mycelial growth does not continue to the bottom.	Excessive grain moisture.	Adjust proper moisture level.
4.	Mycelial growth very thin, hardly penetrates the grains.	Grains too dry.	Adjust proper moisture level.
5.	Mycelia do not grow through substrate or patchy growth.	Grains contaminated with bacteria due to improper sterilization. Less vigorous strain.	Use recommended sterilization time. Use vigorous strain.
6.	Contamination appears on the surface of the grain or on the mycelia plug which was inoculated.	Contamination occurred during inoculation. Mycelia plug or culture contaminated.	Inoculation should be performed in a more aseptic way and observe complete cleanliness.
7.	Mycelia growing very slowly.	Unsuitable substrate. Incubation temperature not suitable. Culture has degenerated.	Use recommended substrate. Check the temperature requirement. Use vigorous culture.

☆ Inoculation is always done near the spirit lamp to avoid contamination.

☆ The working person should wash his/her hands in inoculation chamber using alcohol.

☆ The growth of spawn should be faster and fluffy with whitish in appearance.

☆ Always use well grown mother spawn (15-16 day old).

☆ Mother spawn of beyond 3-4 generation should not be used as it starts degenerating.

☆ PP bag must be marked with firm name, species, quantity, date of inoculation to know the age and type of spawn.

☆ For every new lot of commercial spawn, fresh mother spawn should be used.

☆ Commercial spawn may not be used for further multiplication of seeds as it may lead to higher contamination and decline in yield.

Transport of Spawn

Spawn should not be exposed to temperatures higher than 35°C, while transportation. It is better to pack the spawn bottles or bag in thermocol boxes containing ice. Spawn may be transported during night due to being low temperature. Immediately spawn bags/bottles should not be allowed to stay after shipping and it must be spawned without delay in fresh condition. The spawn can be stored at 5-10°C for one month, in case of redundancy. Spawn of *Volvariella* mushroom should never be refrigerated.

Chapter 9

Techniques of Compost and Casing Preparation

Compost

A mixture of decomposed organic and inorganic substances with nutrient composition selective for the growth and fructification of the common cultivated mushroom. Unlike other traditional crops, soil is not the appropriate substrate for mushroom cultivation. Rather, the substrate for mushroom called compost, is prepared from agro-wastes like straw, stem, shoot, apices *etc.* with organic manure.

Composting Preparation

There are two methods of composting prevalent in India.

The substrate on which button mushroom grows is mainly prepared from a mixture of plant wastes (cereal straw/sugarcane bagasse *etc.*), salts (urea, superphosphate/gypsum *etc.*), supplements (rice bran/wheat bran) and water. In order to produce 1 kg of mushroom, 220 g. of dry substrate materials are required. It is recommended that each ton of compost should contain 6.6 kg nitrogen, 2.0 kg phosphate and 5.0 kg of potassium (N:P:K- 33:10:25) which would get converted into 1.98 per cent N, 0.62 per cent P and 1.5 per cent K on a dry weight basis. The ratio of C:N in a good substrate should be 25-30:1 at the time of staking and 16-17:1 in the case of final compost.

(A) Short Method of Composting

Compost prepared by short method is superior in terms of production quality and is very less prone to infection and disease.

(1) First Phase (Outdoor Composting)

Ingredients

Sl.No.	Ingredients	Quantity
1.	Wheat straw	1000 kg
2.	Chicken manure	600 kg
3.	Urea	15 kg
4.	Wheat bran	60 kg
5.	Gypsum	50 kg

This phase starts with the wetting of raw materials.

1. Wheat straw and chicken manure are mixed thoroughly and placed in layers and sufficient water is added to make stack (almost 5 feet high, 5 feet wide and of any length can be made with the help of wooden boards).
2. On the 2nd day, the stack is turned first time and watered again.
3. On the 4th day the stack is turned second time after adding urea, gypsum, wheat bran, before making watered.
4. On the 8th day the third and final turning is done. The colour of the compost turns dark brown and it starts smelling like ammonia.
5. On the 10th day, the compost is transferred to the pasteurization tunnel having height of 7'.

(2) Second Phase (Indoor Composting)

This is the pasteurization procedure which is done in a closed environment using microbe mediated fermentation process. The whole process is carried out inside a steaming room where an air temperature of 60°C is maintained for 4 hours.

The phase II process is completed in three stages:

(i) Pre-Peak Heat Stage

After 12-15 hours of compost filling, the temperature of compost starts rising and once it arrives 48-50°C, it should be maintained for 36-40 hours with ventilation system. Normally increase in temperature occurs due to self-generation of heat by the compost mass without steam injection.

(ii) Peak Heat Stage

The temperature of compost is raised upto 57-58°C by self-generation of heat from microbial activity. If it is not obtained, live steam is injected in the bulk chamber and maintained for 8 hours to ensure effective pasteurization. Fresh air is introduced by opening fresh air damper up to 1/6 or 1/4 of its capacity and also, air outlet is opened to the same extent.

(iii) Post-Peak Heat Stage

Temperature is brought down gradually up to 48-52°C and is maintained, till no traces of ammonia is detected in compost. This might take 3-4 days in a balanced formulation. When the compost is free from ammonia, full fresh air is introduced by opening damper to its maximum capacity and allow to cool down by bringing the temperature at 25°C, which is considered as the favourable temperature for spawning.

Compost able for spawning should possess 70 per cent moisture, pH 7.2-7.5, ammonia below 0.006 per cent, Nitrogen around 2.5 per cent, dark brown colour and fire fungus (*Actinomycetes*) for good growth.

B. Long Method of Composting

It is primitive method of composting and usually practised in areas where facilities for steam pasteurization are not available. It takes 28 days to make it complete and requires seven turnings. The disadvantage of this method is that, it gives low yield and invites several pest and diseases. The following materials are required for long method of compost.

Sl.No.	Ingredients	Quantity
1.	Wheat straw	300 kg
2.	Wheat bran	15 kg
3.	Ammonium sulphate or calcium ammonium nitrate	9 kg
4.	Super phosphate	3 kg
5.	Murate of potash	3 kg
6.	Urea	3 kg
7.	Gypsum	30 kg
8.	Fipronil 0.3 per cent GR	100 g
9	Follidal or Fenvelrate	100 g

☆ **Day 0:** At this stage, wheat straw with 75 per cent humidity is added with wheat bran, calcium ammonium nitrate, urea, murate of potash, and super phosphate and is mixed. The resulting material is then piled into 1.5 m thickness x1.25 m height with the help of wooden rectangular block. Once the entire material has been stacked up or piled up, water is sprayed twice or thrice to keep the substrate moist. Temperature should be maintained in the range of 70-75°C.

☆ **Day 1 to 5:** Monitoring of the temperature and moisture is required. It should start rising within 24-48 hours after stacking and can reach up to 65-70°C in the central portion.

☆ **Day 6 (1ˢᵗ turning):** On the sixth day first turning is given to the stack to avail even distribution of water and aeration. Turning is done by shaking the outer (top most) part and the inner part of the compost, first separately and then mixing it with the help of wooden buckets.

☆ **Day 10 (2ⁿᵈ turning):** Top most part is separated from the lower part of the compost, water is sprayed on the top part. Again both parts are piled up together in such a way that resulting stack should have reverse parts.

☆ **Day 13 (3ʳᵈ turning):** It is also done in the same way as described earlier. Gypsum and fipronil are added and mixed at this stage.

☆ **Day 16 (4ᵗʰ turning):** The same process of turning is followed.

☆ **Day 19 (5ᵗʰ turning):** The same process of turning is followed.

☆ **Day 22 (6ᵗʰ turning):** The compost is turned in the same manner and required quantity of follidal or fenvelrate is added.

☆ **Day 25 (7ᵗʰ turning):** The same process of turning is followed.

☆ **Days 28:** On this day, compost is checked for the smell like ammonia. If no ammonia persists, compost is ready for spawning. Ammonia should not be more than 8-10 ppm at the spawning stage.

Casing Soil

Casing means covering the top surface of compost bag after spawn run is over to stimulate pinning and fruiting of mushrooms. The casing layer keeps the compost surface from drying and work as a water reservoir for the mature mushroom. Thick casing layer of 3-4 cm is usually applied on the surface. Casing provides physical support, moisture and exchange of gases within the surface of the compost which helps in proper growth of the mycelium.

Requisites of Casing

☆ Soil or casing material should have good aeration capacity *i.e.* Texture should be quite porous so that exchange of gases can take place easily.

☆ It should be free from disease organisms, insects and undecomposed organic matter.

☆ The water holding capacity of casing must be higher.

Advantage of Casing

☆ Casing enhances water holding capacity and helps to absorb water quickly but releases it slowly.

☆ It prevents the evaporation of excessive moisture from the beds and after each watering a cool layer is provided in form of casing.

☆ Casing material (soil) conserves moisture and provides humidity as it can hold more water for longer period.

☆ Casing soil encourages to vegetative mycelium for fruiting only when it spreads into the medium which is deficient in nutrition.

☆ Casing maintains humidity and temperature in cropping room evaporative cooling.

Material of Casing

Various materials have been tested as a casing substrate throughout the world by different workers. Throughout the world, peat is the ideal casing material because of its high water holding capacity and porosity. However, in India it is not available in plenty and is also not good quality, and well decomposed farm yard manure, well decomposed spent mushroom compost, decomposed coir pith alone or in different combinations among themselves or with burnt rice husk or garden soil etc are popular in India. In India, casing soil is prepared using following ingredients as the substitutes:

Mixture of Substrates	Ratio
Two years old manure + garden soil	3:1
Two year old manure + garden soil	2:1
Two year old manure + spent compost	1:1
Two year old manure + spent compost	2:1
Two year old manure + spent compost	1:2
Garden soil + sand mixture	4:1

Preparation of Casing Soil

The casing soil is piled on cemented floor. Thorough turning of the mixture is done and it is covered with polythene sheet for next 3- 4 days. The pH of the mixture is tested and adjusted to a value of 7.5-7.8. Pasteurization of casing soil as heat treatment at 65°C for 6-8 hours is found to be much more effective. On small farms, 4 per cent solution of formalin or chloropicrin is usually applied for sterilization. Casing soil of thick layer *i.e.* 3-4 cm is spread uniformly on the compost when the surface gets covered by white mycelium of the mushroom fungus. Formalin solution (0.5 per cent) is sprayed thereafter. Temperature and humidity of the crop room should be maintained at 14-18°C and 80-85 per cent, respectively. Proper ventilation should be arranged with water being sprayed once or twice a day.

Spawned Casing

Mac Canna and Flanagan in late 1960's discovered the spawned casing technique. spawn run compost when added to casing reduces the time taken for formation of pin head and increase the yield also. Casing added with already spawn run compost is an improved agronomic practice and helps in mycelial growth and mushroom yield.

Care Should be taken during Casing

☆ Casing material should not be sieved but used as such with clumps, which permits more air spaces in casing.

☆ Top casing surface should have small mounts and valleys.

☆ Care should be taken to prevent re-infection of the casing materials.

1. Vermicompost

2. Farm Yard Manure

3. Garden Soil

Figure 9.1: Substrates for Preparation of Casing Soil.

☆ Store casing material in a sterilized/clean room before use, in polythene bag or synthetic cloth bags.

☆ Apply water to casing in a few instalments so that water does not run into spawn run compost.

Chapter 10

Techniques of Spawning

Spawning

Spawning is the inoculation of the mushroom culture into the substrate or compost. It is the actual planting of the spawn and requires care depending on the species of mushroom and methods being followed. After spawning spawn-run stage comes where the mycelium of the fungus is allowed to grow. For spawning, completely colonized and fresh spawn should be used.

Spawning of Oyster Mushroom

When the pasteurized substrate had cooled down to room temperature, it was ready for filling and spawning. At this stage, substrate moisture content was about 70 per cent. Polythene bags (35 x 50 cm, 150 gauges) or polypropylene bags (35x50 cm, 80 gauges) are used for its cultivation. Freshly prepared (15-18 days old) grain spawn is best for spawning. The spawning should be done in a pre-fumigated room (48 hour with 2 per cent formaldehyde). Spawning can be done in layer spawning or mixed.

Layer Spawning

In case of layer spawning, substrate is filled in bag, pressed to a depth of 8-10 cm and broadcast with a handful of spawn above it. Similarly, 2nd and 3rd layers of substrate are put and simultaneously after spawning, the bags were closed for spawn running (Figure 10.1a).

Mixed Spawning

In mixed spawning, pasteurized straw is mixed with spawn 2 to 3 per cent of the dry wt. of the substrate and filled in bags (Figure 10.1b). One bottle spawn of 300 g is sufficient for 10-12 kg of wet substrate or 2.8 to 3 kg of dry substrate. After that it is gently pressed, and the bags are sealed for spawn running.

a. Layer Spawning   b. Mixed Spawning

Figure 10.1a-b: Spawning of Oyster Mushroom.

10 to 15 small holes (0.5-1.0 cm dia) should be made on all sides especially in the bottom for leaching of excess water. Spawned bags are stacked on racks in neat and clean place, in closed position. Temperature at 25-35°C and humidity at 85-90 per cent is maintained by spraying water twice a day on walls and floor. It takes 15-20 days when bags are fully covered with white and pink mycelium respectively.

Spawning of Button Mushroom

Three methods of spawning were employed for compost inoculation.

(1) Surface/Broadcast Spawning

Compost is filled in formalin sterilized wooden boxes (60-90 x 15–23 cm, accommodating 6-10 kg compost) and the spawn evenly spread over the surface of the box and allowed to grow into the compost. The top portion is covered with a thin layer of compost and covered with formalin sterilized newspaper sheet to prevent loss of moisture content in mushroom beds. Chief advantages of the broadcast method are convenience, labour saving, and time saving. Its disadvantages include exposure of grains to drying conditions, a slow start due to dry surface conditions, and the time required for mycelium to grow throughout the full depth of compost and establish total colonization -thereby allowing competitors more time to gain access to the compost in the lower portion of the bed. Also, the nutritional value of the grain is lost to the crop. Broadcast spawning is rarely practised in the mushroom industry today.

(2) Layer Spawning

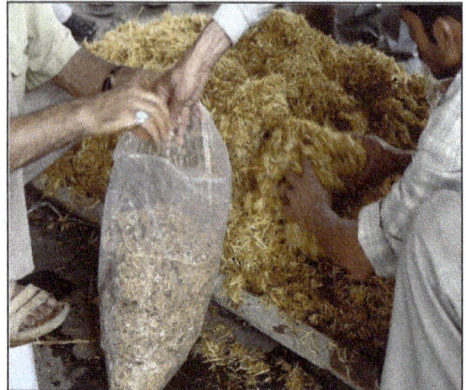

Spawning is done by scattering the spawn on boxes beds once when half-filled with compost and secondly, after the complete filling of the boxes. The spawn is gently pressed down with the forefinger uniformly each time and trays are covered with formalin sterilized newspaper sheets.

(3) Mixed Spawning

The spawn is mixed throughout the compost @ 500 gm/100kg of compost and filled in formalin sterilized wooden boxes and covered with formalin sterilized newspaper. Advantages to mixed spawning are it reduces the spawn growing period, spawn is evenly distributed in the compost and increases mushroom yield.

Spawn Run

The process of colonization of compost from grain inoculum is called spawn-run.

These mushroom boxes are placed in growing chamber, where temperature ranges between 22-28°C and 90-95 per cent RH (relative humidity). The paper over the beds is sprayed regularly with water to prevent drying out and humidity is built up by frequently watering the floor and walls. The room was kept closed as only a small amount of fresh air re-circulating within the crop room for maintaining the carbon dioxide levels. Mushroom boxes are completely colonized by mushroom mycelium within 20-31 days. The compost becomes lighter in colour and the mycelium is seen as thin white-threads.

Care which should be taken during Spawning

- ☆ Use of freshly prepared pure culture spawn
- ☆ Spawning under aseptic conditions.
- ☆ Proper treatment of spawning area and tools with formalin, while hands with dettol.
- ☆ Maintaining good hygienic conditions during spawning by keeping all the doors/windows closed.

Chapter 11

Cultivation Technique of Button Mushroom (*Agaricus* sp.)

Button mushroom is the most popular of the cultivated mushrooms. This mushroom is extensively cultivated throughout the world and contributes about 40 per cent of the total world production of mushroom. This is the first mushroom to be commercially exploited on industrial scale. The genus *Agaricus* has two cultivated species namely *A. bisporus* (temperate button mushroom) and *A. bitorquis* (tropical or high temperature tolerant white button mushroom).

Figure 11.1: button Mushroom.

Nomenclature

Button mushroom has a complicated taxonomic history. It was first described by English botanist Mordecai Cubitt Cooke in his book *"Handbook of British Fungi"* in 1871 as a variety (var. *hortensis*) of *Agaricus campestris*. Jakob Emanuel Lange, Danish mycologist later reviewed a cultivar specimen, and dubbed it *Psalliota hortensis* var. *bispora* in 1926. In 1938, it was promoted to status of species and renamed *Psalliota bispora*. Emil Imbach imparted the current scientific name of the species, *Agaricus bisporus* in 1946. The specific epithet *bispora* distinguishes the two-spore basidia from four-spore varieties.

Kingdom: Fungi

Phylum: Basidiomycota

Class: Agaricomycetes

Order: Agaricales

Family: Agaricaceae

Genus: *Agaricus*

Species: *A. bisporus*

Other important cultivated species: *A. bitorquis* and *A. brunnescens*

Agro-Climatic Requirements

☆ Best period for cultivation: Winter months (October to February) for plains of India.

☆ Required optimum temperature: *A. bisporus* requires 20-22°C for spawn run and 16-18°C for fruiting. However another species *A. bitorquis* needs higher temperature, 28-30°C for spawn run and 25°C for fruiting.

☆ Relative humidity (RH): 80-90 per cent

☆ The major producing states are Himachal Pradesh, Uttar Pradesh, Punjab, Haryana, Jammu and Kashmir, Maharashtra, Andhra Pradesh, Tamil Nadu and Karnataka.

Requirements

The require materials are listed below cultivation of button mushroom.

Sl.No.	Raw Materials	Sl.No.	Appliances/Tools
1.	Fresh spawn (15 days old)	1.	Mushroom house
2.	Substrate (wheat/maize/pulse straw/sawdust)	2.	Water tank
3.	Poly propylene bag (PP bag)/trays (mostly wooden trays 1x1/2 m.	3.	Casing preparation floor
4.	Compost	4.	Cemented floor for drying of straw
5.	Casing	5.	LPG/Electric stove
6.	Rubber band	6.	Thermometer and hygrometer
7.	Chemicals (Bavistin and Formalin)		

Cultivation Technique

The cultivation of white button mushroom requires more ventures as it take more time, cost and space to create conducive situation and commercial production. Cultivation of this mushroom takes place at 15-17°C, therefore it requires additional cooling system. More over it also requires controlled cropping rooms and infrastructure for compost and casing preparation. Button mushroom cultivation can be carried out through following straight forward steps:

1. Procurement of spawn
2. Compost preparation
3. Spawning
4. Casing
5. Cropping and harvesting
6. Post harvest management

1. Procurement of Spawn

Fresh *A. bisporus* spawn should be about 15 days old when mycelium has formed complete coating around the grain is taken. The spawn must be procurement from reliable sources because the purity of spawn is an important parameter for mushroom cultivation. It is advisable to purchase or procure the spawn from institutes or registered companies **(The process of spawn preparation is already mentioned in Chapter 8)**.

2. Compost Preparation

A mixture of decomposed organic and inorganic substances with nutrient composition selective for the growth and fructification of the common cultivated mushroom. Unlike other traditional crops, soil is not the appropriate substrate for mushroom cultivation. Rather, the substrate for mushroom called compost, is prepared from agro-wastes like straw, stem, shoot, *etc.* with organic manure. The substrate on which button mushroom grows is mainly prepared from a mixture of plant wastes (cereal straw/sugarcane bagasse *etc.*), salts (urea, superphosphate/ gypsum *etc.*), supplements (rice/wheat bran) and water **(The process of composting preparation is already discussed in Chapter 9)**.

3. Spawning

Spawning is done by layer methods @ 4-5 per cent of wet weight of the substrate. Prepared compost is spawned @ 500 to 750 g/100 kg compost by layer method **(The process of spawning is already discussed in Chapter 10 with detail)**. After spawning, the compost is filled in polythene bags (90x90 cm, 150 gauge thick having a capacity of 20-25 kg/bag)/trays (mostly wooden trays 1x1/2 meter accommodating 20-30 kg. compost) set on shelves/rack with sufficient space to allow ease of air circulation. The bags or tray are usually made covered with formalin treated news paper. If it is bag, it should be folded at the top and covered up. After spawning, temperature and relative humidity of crop room should be maintained at 18-22°C and 85-90 per cent, respectively. Water should be sprayed on bags over the covered news papers, walls and floors of the crop room.

4. Casing

After 12-14 days the compost beds surface is covered by white mycelium of the mushroom should be covered with a 3-4 cm thick layer of casing soil to induce pinning and fruiting (**for detail refer Chapter 9**). Formalin solution (0.5 per cent) is then being sprayed to avoid other microbial contamination. Before application casing soil should be either pasteurized (at 66-70°C temperature for 7-8 hours) or treated with formaldehyde @ 2 per cent and bavistin @ 2 gm/liter or steam sterilized. The treatment needs to be done at least 15 days before the material is used for casing.

5. Pin Head Initiation and Fruiting

Under favourable environmental conditions *viz.* temperature (initially 23±2°C for about a week and then 16-18°C), relative humidity (above 85 per cent), proper ventilation and CO_2 concentration (0.08-0.15 per cent) pin head initiation takes place after 10-12 days, which start growing and gradually develop into button stage (fruiting bodies).

6. Harvesting

Fruiting bodies (button stage 2.5 to 4 cm) of the mushroom can be harvested for around 50-60 days. The crops should be harvested before the gills open as this may decrease its quality and market value. To harvest, the base of the mushroom is grasped with the help of stem and rocked it free from the compost, avoiding pulling which might damages the mycelia for further crops. Once the harvesting is complete, the gaps in the beds should be filled with fresh sterilized casing material and then watered. About 10-14 kg fresh mushrooms per 100 kg. fresh compost can be obtained in two months crop. Short method used for preparation of compost under natural conditions gives more yield (15-20 kg/100 kg compost).

Figure 11.2: Compost Covered Formalin Treated News Papers.

1. Mycelial growth of *A. bisporus*

2. Casing followed by pin head (stage I) initiation

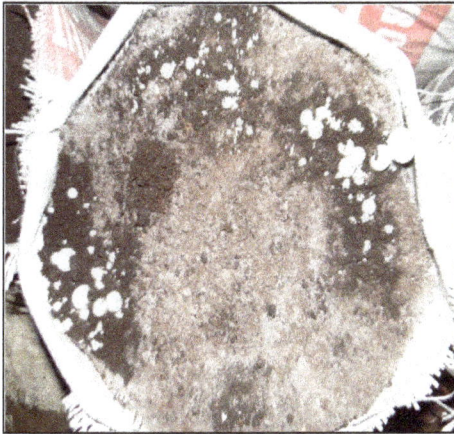

3. Pin head initiation (stage II)

4. Fruiting of button mushroom

Figure 11.3: Different Stages during Growth and Fruiting of Button Mushroom.

6. Post Harvest Management

The button mushrooms are best when it is consumed as fresh. Storage in refrigerator for a few days is possible if they are placed between moist paper towel or damp cloth. Harvested mushrooms are stored in polythene bags at 4-5°C up to 3-4 days. The mushrooms are usually packed in unlabelled simple polythene or polypropylene for retail sale. Bulk packaging does not exist. In developed countries, modified atmosphere packaging (MAP) and controlled atmosphere packaging (CAP) are in trend.

Chapter 12

Cultivation Technique of Oyster Mushroom (*Pleurotus* sp.)

Pleurotus mushroom which is generally known as 'oyster mushroom' globally, but it is called "Dhingree" in India. The oyster mushroom is one of the most suitable fungus for converting protein rich food utilizing various agro wastes without composting. The fruit bodies of this mushroom are fan or spatula shaped with different shades of white, cream, grey, yellow, pink or light brown depending upon the species. It has its origin from Greek word 'pleuro' which means formed laterally or in side way position, particularly refers to the lateral position of the stipe (stem) in relation to the pileus (cap). Different species of *Pleurotus* mushroom which are commercially produced in both India and overseas listed in Table 12.1:

Table 12.1: Temperature Range, Fruit Colour and Growth Pattern of different Species of Oyster Mushroom.

Sl.No.	Scientific Name	Common Name	Optimum Temperature Range (ºC)	Colour	Growth pattern
1.	*Pleurotus ostreatus*	Black oyster mushroom	20-22	Black	Vertically long bunches
2.	*Pleurotus florida*	White mushroom	20-28	White	bunches
3.	*Pleurotus sajor-caju*	Grey mushroom	22-30	Dark grey	Fan shaped
4.	*Pleurotus d'jamour*	Pink mushroom	22-26	Pink	Gracious
5.	*Pleurotus cornicopiae*	Golden mushroom	20-22	–	–
6.	*Pleurotus eryngi*	King oyster	18-22	–	–

Figure 12.1: *Pleurotus* **Mushroom.**

Nomenclature

The term *Pleurotus* coined by Paul Kummer (1887) has been derived from two Greek words: *pleure* means "side" and *ous* means "ear". The taxonomic position of *Pleurotus* is given below.

Kingdom: Fungi

Phylum: Basidiomycota

Class: Agaricomycetes

Order: Agaricales

Family: Pleurotaceae

Genus: Pleurotus

Species: P. ostreatus

Other cultivated species: *P. sajorcaju, P. florida, P. eryngii* and *P. djamor*

Agro-climatic Requirements

☆ September to March is considered as the suitable period for best cultivation.

☆ Optimum temperature for growth: The optimum for the growth of most of the species of *Pleurotus* are 20-28°C. However one of the species, *P. sajor-caju* can grow up to 30°C temperature.

☆ Relative humidity (RH): 80-85 per cent.

☆ The major producing states are Odisha, Karnataka, Maharashtra, Andhra Pradesh, Madhya Pradesh, West Bengal and in the North-Eastern States of Meghalaya, Tripura Manipur, Mizoram and Assam.

Requirements

The require materials are listed below for cultivation of oyster mushroom

Sl.No.	Raw Materials	Sl.No.	Appliances/Tools
1.	Fresh spawn (15 days old)	1.	Mushroom house with bamboo racks
2.	Substrate (wheat/maize/pulse straw/sawdust)	2.	Water tank
3.	Poly propylene bag (size-35 x 50 cm, 80 gauges)	3.	Water bath
4.	Supplement (rice/wheat bran)	4.	Cemented floor for drying
5.	Rubber band	5.	LPG/Electric stove
6.	Chemicals (Bavistin and formalin)	6.	Fry pan (20-25 kg) for boiling substrate
		7.	Mesh (for filter water from substrate-3x6 fit
		8.	Thermometer and hygrometer

Cultivation Technique

The cultivation techniques of every species of oyster mushroom are more or less similar. The most economic and farmers friendly technique is being dealt here which consist of following steps:

1. Procurement of spawn
2. Substrate preparation
3. Sterilization/pasteurization of substrate
4. Spawning
5. Cropping and harvest
6. Post harvest management

1. Procurement of Spawn

Fresh *Pleurotus* spawn should be about 15 days old when mycelium has formed complete coating around the grain is taken. The spawn must be procured from reliable sources because the purity of spawn is an important parameter for mushroom cultivation. It is advisable to purchase or procure the spawn from institutes or registered companies **(The process of spawn preparation is already mentioned in Chapter 8).**

2. Substrate Preparation

Oyster mushroom was grown on various substrates *viz.*, Wheat straw, paddy straw, pulse straw, sawdust, maize stalk and leaves of maize, millets and cotton *etc.* It can also be cultivated by using industrial wastes like paper mill sludges, coffee by-products, tobacco waste *etc.* Since wheat straw is easily available and cheap, hence it is widely used. Wheat straw used are fresh and well dried. Substrate are soaked in fresh water in water tank with formalin @ 1 ml/litter and bavistin @ 2 gm/liter and left overnight. Excess water from straw is drained off by spreading it on cemented floor in shade.

3. Sterilization/Pasteurization of Substrate

Water is boiled in a wide mouth container such as tub or drum. The wet substrate are filled in gunny bags. The filled bag is dipped in hot water of 80-85°C for about 15-20 minutes. To avoid floating, it is pressed with some heavy material or wooden piece. After pasteurization, excess hot water is drained off from container so that it can be reused for other sets and hot water temperature is maintained at 80-85°C for all sets to achieve pasteurization. During this process, possible fungi and bacteria in substrate mixture are killed. Pasteurized substrate is cooled down at room temperature.

4. Spawning

After cooling, the pasteurized substrate are mixed with pasteurized rice/wheat bran @ 6 per cent of wet weight of the substrate. Resulting mixture is filled

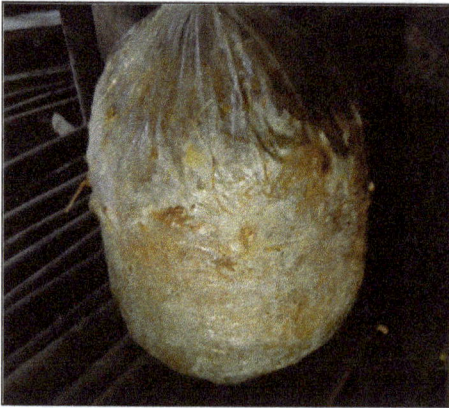

1. Fully impregnation of bag with white mycelium

2. Pinhead head (stage I) initiation

3. Pinhead head (stage II) initiation

4. Complete growth of oyster mushroom as fruit

Figure 12.2: Different Stages during Growth and Fruiting of Oyster Mushroom.

in polypropylene bag (size-35 x 50 cm, 80 gauges). During spawning, substrate moisture content should be about 70 per cent. The mixture is spawned (using grain spawn) @ 3 per cent (wet weight basis) under aseptic conditions. 4-5 kg spawn is sufficient for 100 kg prepared substrate. Spawning can be done by two methods: 1) layer spawning 2) Mixed spawning **(The process of spawning is already mentioned in Chapter 10).** After spawning the bags are tied on the top with the help of rubber bands and holes of about 1 cm diameter are made at 10-15 cm distance across the surface for free diffusion of gases and heat generated inside. The spawned bag is shifted into clean, close and dark room and maintained at 24-25°C and 80-85 per cent relative humidity. RH may be maintained by spraying water twice a day on walls and floor. The spawn run takes about 12-14 days.

5. Cropping and Harvest

After 20-22 days, when bags are fully impregnated with white mycelium, it is shifted to cropping room. The open bags are kept on racks 20 cm apart at 50-60 cm distanced shelves. Relative humidity is maintained by spraying water twice a day on the walls and floor of the room. A light spray of water is given on bags as soon as the small pin heads emerge. Mushroom primordial begin to form, usually after 3 to 4 days of opening of bags. Matured mushrooms are ready to harvest in another 2 to 3 days. The fruit of mushroom should be harvested by twisting before release of spores, so that the stubs are not left on the beds (straw). It is advisable to pick mushrooms by hand instead of using knife, at the same time from a whole cube, so that next flush may come out synchronously. After 1st flush of harvest, 0.5 to 1 cm outer layer of the block gets scrapped. This helps to initiate 2nd flush which appears after 10 days. Depending upon the conditions, 2-3 flushes can be produced within about 45 days.

6. Post Harvest Management

Harvested fresh mushrooms should be packed in perforated (5-6 tiny holes) polythene bags for marketing. For transporting the fruit bodies are stacked in trays or baskets covered with thin polythene sheet with perforation. Freshly harvested mushrooms can be stored at low temperature (0-5°C) for 6-7 days without losing quality. Oyster mushroom can be shed dried for 2 days and dried product marketed in polythene bags. Dried mushrooms were soaked in water for 10 minutes before use.

Chapter 13

Cultivation Technique of White Milky Mushroom (*Calocybe indica*)

Milky mushroom (*Calocybe indica*) is robust, fleshy, milky white with a long stipe and a small pileus, umbrella like, which resembles button mushroom. Milky mushroom is also known as dudh chhata in West Bengal. The major advantage of this mushroom is that it can be best fitted in relay cropping when no other mushroom can be grown at higher temperature. This mushroom has a superior shelf life 3-4 days without loss of colour and appearance, and can be transported to short distances for marketing without damaging its quality. Now days this mushroom is very popular due to its longer shelf life and adaptable to warm and humid conditions.

Figure 13: Milky Mushroom (*Calocybe indica*).

Nomenclature

The term *Calocybe* has been derived from two Greek words: *kalos* means "pretty" and *cubos* means "head". The taxonomic position of *Pleurotus* is given below.

Kingdom: Fungi

Phylum: Basidiomycota

Class: Agaricomycetes

Order: Agaricales

Family: Lyophyllaceae

Genus: Calocybe

Species: Calocybe indica

Type species: Calocybe gambosa

Agro-Climatic Requirements

☆ Best period for cultivation are summer months.

☆ Temperature for vegetative growth (spawn run): 26-28°C and 30-35°C for fruiting.

☆ Relative humidity (RH): 85-90 per cent.

☆ The major producing states are Tamil Nadu, Kerala, Karnataka, Odisha, Haryana and West Bengal.

Requirement

The require materials are listed below for cultivation technique of white milky mushroom.

Sl.No.	Raw Materials	Sl.No.	Appliances/Tools
1.	Fresh spawn (15 days old)	1.	Crop room/Hut- bamboo platform
2.	Substrate (dried/wheat/paddy straw)	2.	Water tank
3.	Wheat bran	3.	Cemented floor for drying of straw
4.	PP bag (60 x 30 cm size)	4.	LPG/Electric stove
5.	Rubber band	5.	Casing preparation floor
6.	Chemicals (Bavistin and formalin)	6.	Mesh (for filter water from substrate-3x6 fit
		7.	Thermometer and hygrometer

Cultivation Technique

The entire technique of white milky mushroom cultivation can be done through following steps:

1. Procurement of spawn
2. Substrate preparation
3. Sterilization/pasteurization of substrate

4. Spawning
5. Cropping and harvest
6. Post harvest management

1. Procurement of Spawn

Fresh *Calocybe* sp. spawn should be about 15 days old when mycelium has formed complete coating around the grain is taken. The spawn must be procured from reliable sources because the purity of spawn is an important parameter for mushroom cultivation. It is advisable to purchase or procure the spawn from institutes or registered companies **(The process of spawn preparation is already mentioned in Chapter 8).**

2. Substrate Preparation

Milky mushroom (*Calocybe indica*) can be grown on wide range of substrates as in case of oyster mushroom. It can be grown on substrates containing lignin, cellulose and hemicelluloses. Cereal straw (paddy/wheat) easily available in abundance, is being widely used. Fresh dried substrate is most suitable for its cultivation. Substrate are soaked in fresh water in water tank with formalin @ 1 ml/litter and bavistin @ 2 gm/liter and left overnight. Excess water from straw is drained off by spreading it on cemented floor under shade.

3. Sterilization/Pasteurization of Substrate

Water is boiled in a wide mouth container such as tub or drum. The wet substrate was filled in gunny bags. The filled bag was dipped in hot water of 80-85°C for about 15-20 minutes. To avoid floating, it is pressed with some heavy material or wooden piece. After pasteurization, excess hot water is drained off from container so that it can be re-used for other sets and hot water temperature was maintained at 80-85°C for all sets to achieve pasteurization. During this process, possible fungi and bacteria in substrate mixture are killed. Pasteurized substrate is cooled down at room temperature

4. Spawning

After cooling, the pasteurized substrate are mixed with pasteurized rice/ wheat bran @ 6 per cent of wet weight of the substrate. Resulting mixture is filled in polypropylene bag (size-35 x 50 cm, 80 gauges). During spawning, substrate moisture content was about 70 per cent. Spawning is done by layer methods @ 3-4 per cent of wet weight of the substrate **(The process of spawning is already discussed in Chapter 10).** The spawned bag is shifted to the spawn running room, in closed position and maintained at 25-28°C and 80-85 per cent relative humidity (RH). RH is maintained by spraying water twice a day on walls and floor. The spawn run takes about 15-16 days.

5. Casing

Casing soil is applied to about 2-3 cm height on the top of the open bed surface after 15 to 16 days of spawning when the substrate gets fully colonized by the white fungus mycelium **(for detail refer Chapter 9)**. After application of casing, light spray of water is done regularly to maintain casing in wet state.

6. Cropping and Harvest

After casing the bags are shifted to cropping room where temperature range 30-35°C and 75-80 per cent relative humidity are maintained and the crop begins to appear in another 7-10 days after complete case run. The primordial are developed into fully grown harvestable mushrooms in 3-4 days. Mushrooms are harvested after attaining the height of 7-8 cm by twisting. The complete cycle takes about 45 days on synthetic logs. One kg of mushroom gives rise 10-12 fruiting bodies, yielding crop (500-600 g/kg dry weight substrate).

7. Post Harvest Management

Harvested mushroom can be cleaned and packed in perforated polythene/ polypropylene bags for marketing. Mushrooms can also be wrapped in kiln film for longer storage. Milky mushroom has a superior shelf life, and can be transported to short distances for marketing without damage to its quality.

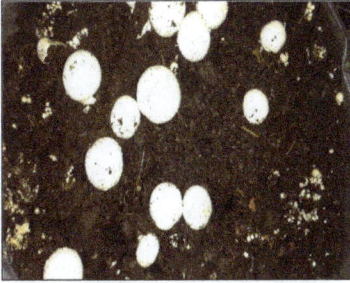

Agaricus bisporus

Pleurotus sp.

Calocybe indica

Volvariella volvacea

Figure 2.1: Major Cultivated Species of Edible Mushroom. (p. 14)

1. Vermicompost

2. Farm Yard Manure

3. Garden Soil

Figure 9.1: Substrates for Preparation of Casing Soil. (p. 62)

a. Layer Spawning   b. Mixed Spawning

Figure 10.1a-b: Spawning of Oyster Mushroom. (p. 64)

Figure 11.1: button Mushroom. (p. 67)

1. Mycelial growth of   2. Casing followed by pin
A. bisporus   head (stage I) initiation

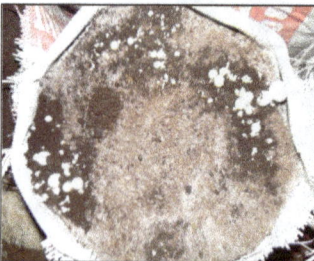

3. Pin head initiation (stage II) 4. Fruiting of button mushroom

Figure 11.3: Different Stages during Growth and Fruiting of Button Mushroom. (p. 71)

Figure 12.1: *Pleurotus* Mushroom. (p. 74)

1. Fully impregnation of bag with white mycelium

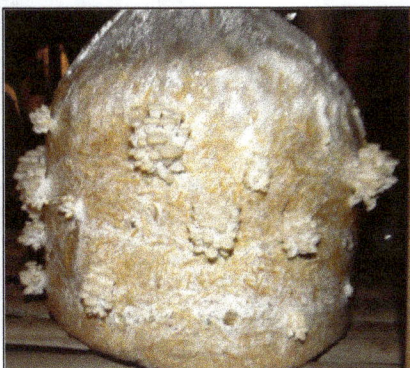

2. Pinhead head (stage I) initiation

3. Pinhead head (stage II) initiation

4. Complete growth of oyster mushroom as fruit

Figure 12.2: Different Stages during Growth and Fruiting of Oyster Mushroom. (p. 76)

Figure 13: Milky Mushroom (*Calocybe indica*). (p. 79)

Figure 14.1: Paddy Straw Mushroom. (p. 83)

Figure14.2: Spawning at Paddy Straw Bundle. (p. 86)

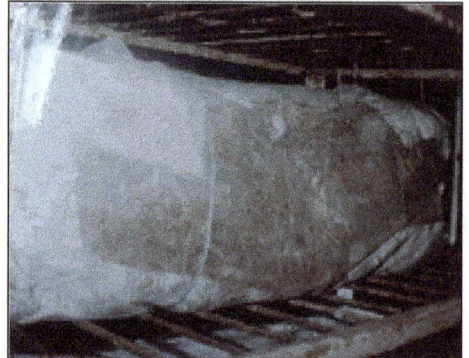

Figure14.3: Paddy Straw Bundle Wrapped with Poly Propylene Bag. (p. 86)

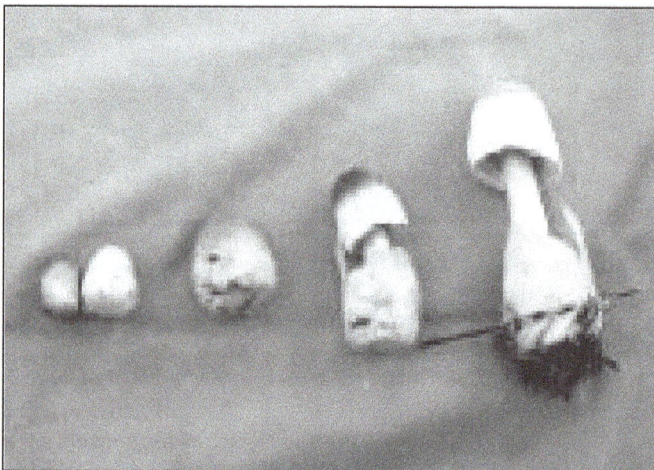

Figure 14.4: Different Growth Stages of Fruiting Bodies of the *Volvariella volvacea* grown on Paddy Straw. (p. 86)

Figure 16.1: Shiitake Mushroom (*Lentinula edodes*) on Saw Dust. (p. 94)

Figure 16.2: Fruiting Body of *Ganoderma lucidum*. (p. 96)

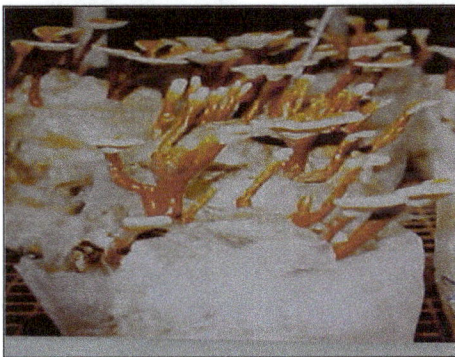

**Figure 16.3: Different Stages during Growth and Fruiting of
Ganoderma lucidum on Saw Dust. (p. 98)**

1. *Amanita phalloides*

2. *Amanita* sp.

3. *Ganoderna* sp.

4. *Microporous* sp.

Figure 17.1: Few Major Species of Poisonous Mushroom. (p. 103)

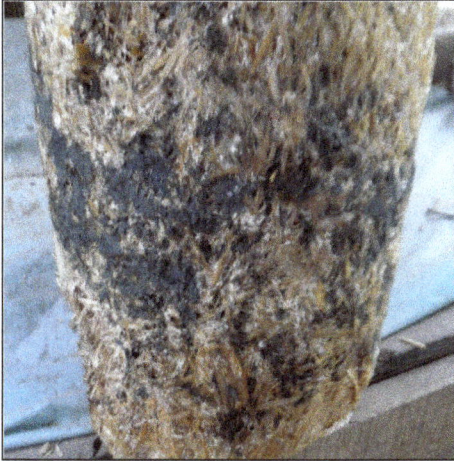

1. Green mould (*Trichoderma* sp.) on mushroom bag

2. Green mould (*Trichoderma* sp.) in spawn bag

3. Dry Bubble (*Verticillium fungicola*)

4. Wet bubble (*Mycogyne perniciosa*)

5. White plaster moulds (*Scopulariopsis fumicola*)

6. Bacterial yellow blotch (*Pseudomonas tolassi*)

Figure 20.1: Diseases of Mushroom. (p. 122)

Figure 23.1: Arrangement of Mushroom Bags in the Racks/Shelves of Bamboo in Mushroom House. (p. 140)

Figure 23.2: Arrangement of Mushroom Bags in the Racks/Shelves of Cement in Mushroom House. (p. 140)

Figure 23.3: Arrangement of Mushroom Bags in the Air Space in Mushroom House. (p. 140)

Figure 23.4: Artificial Rain Water on Mushroom Bags. (p. 140)

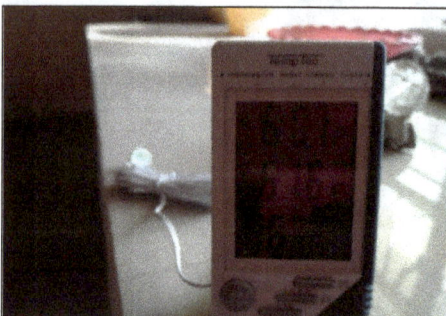

Figure 23.5: Hygrometer Device which Observe the RH in the Mushroom House. (p. 140)

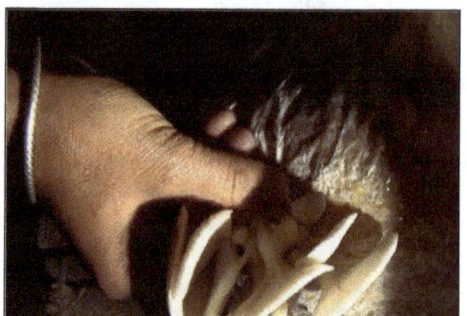

Figure 23.6: Harvesting and Picking of Mushroom by Hand with Twisting Motion. (p. 144)

Chapter 14

Cultivation Technique of Paddy Straw Mushroom (*Volvariella volvacea*)

Paddy straw mushroom (*Volvariella volvacea*) also known as Chinese mushroom/ tropical paddy straw mushroom. Sometimes, paddy straw mushroom is also called as "warm mushroom" as it grows at relatively high temperature. Its name derives from the substrate on which it originally was grown. Cultivation of *Volvariella* was believed to have begun in China as early as 1822. It ranks sixth among the cultivated

Figure 14.1: Paddy Straw Mushroom.

mushrooms at the global level and is most popular mushroom in South East Asia. It is very delicious mushroom amongst the edible group having good combinations of all attributes like flavor, aroma, delicacy, high content of protein, vitamins and minerals. Moreover, It is quickly growing fungus, which can be harvested on 12th or 13th day only. But it has poor shelf life, therefore, it is better to either consume within hours of its harvest, or dehydrate it for preservation. Genus *Volvariella* takes its name from 'Volva' means a wrapper; which completely envelops the main fruit body during the young stage.

Nomenclature

The taxonomic position of ***Volvariella*** is given below.

Kingdom: Fungi

Phylum: Basidiomycota

Class: Agaricomycetes

Order: Agaricales

Family: Pluteaceae

Genus: Volvariella

Species: Volvariella volvacea

Agro-Climatic requirements

☆ Best period for cultivation are warmer months.

☆ Temperature for spawn run and cropping 30-35°C.

☆ Relative humidity (RH): 85-90 per cent.

☆ The major producing states are Odisha, Andhra Pradesh, West Bengal, Tamil Nadu and Kerala.

Requirements

The require materials are listed below for cultivation of paddy straw mushroom.

Sl.No.	Raw Materials	Sl.No.	Appliances/Tools
1.	Fresh spawn (15 days old)	1.	Mushroom house with bamboo platform
2.	Substrate (dried paddy straw)	2.	Water tank
3.	Supplement (besan)	3.	Cemented floor for drying of straw
4.	Plastic sutli	4.	LPG/electric stove
5.	Chemicals (Bavistin and formalin)	5.	Thermometer and hygrometer

Cultivation Technique

The technique which are given below for the cultivation of paddy straw mushroom are very economic and farmers friendly. The entire technique of paddy straw mushroom cultivation can be done through following steps:

1. Procurement of spawn
2. Substrate preparation
3. Sterilization/pasteurization of beds
4. Spawning at substrate
5. Cropping and harvest
6. Post harvest management

1. Procurement of Spawn

Fresh *Volvariella* spawn should be about 15 days old when mycelium has formed complete coating around the grain is taken. The spawn must be procured from reliable sources because the purity of spawn is an important parameter for mushroom cultivation. It is advisable to purchase or procure the spawn from institutes or registered companies (**The process of spawn preparation is already mentioned in Chapter 8**).

2. Substrate Preparation

Paddy straw mushroom is grown on limited substrates *viz.*, rice straw, leaves of banana/water hyacinth *etc.* Since rice straw is easily available and cheap, hence it is widely used. Fresh dried paddy straw is most suitable for its cultivation. The straw is tied into bundles of 1.2 m long and of 25 cm diameter. Tied bundles of rice straw steep in water in a cemented tank overnight. The bundles are taken out from tank and put on a cemented floor for few hours to drain out excess water.

3. Sterilization/Pasteurization of Beds

Water is boiled in a wide mouth container such as tub or drum. Prepared bundle dipped in hot water at 80-85°C for about 1 hours. To avoid floating, it is pressed with some heavy material or wooden piece. After pasteurization, the bundles put on a cemented floor for few hours to drain out excess water and cooled down to room temperature. During this process, possible microbes present in the beds are killed.

4. Bed Preparation

The mushroom beds are prepared on a raised bamboo platform inside a thatched hut. Four pre-wetted bundles are placed side by side on this platform, facing all the loose ends on one side. Then another four bundles are placed with their tied ends on the opposite side, so that loose ends of all 8 bundles meet and overlap each other in the middle.

5. Spawning

Spawn is applied on this first layer about 15 cm away from the outer edge. The amount of spawn to be used is calculated at 1.5 per cent of dry weight basis or 0.5 per cent of wet weight basis of the substrate. On top of the spawned first layer, a little quantity of Besan (about 200 g/bed) is applied along with the spawn. Again a second layer of 8 spawn bed is placed in a similar fashion and spawned on top of the spawned first layer of eight bundles. A third layer is again laid on top of the second layer and followed by 4th layer of bundles and spawning. These 32 bundles

Figure14.2: Spawning at Paddy Straw Bundle.

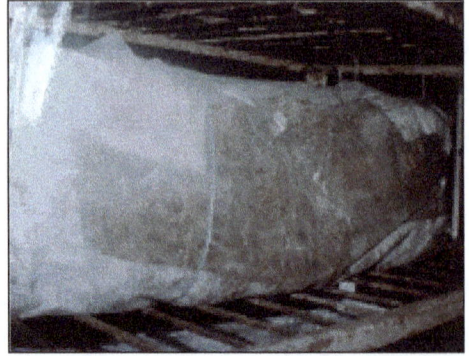

Figure14.3: Paddy Straw Bundle Wrapped with Poly Propylene Bag.

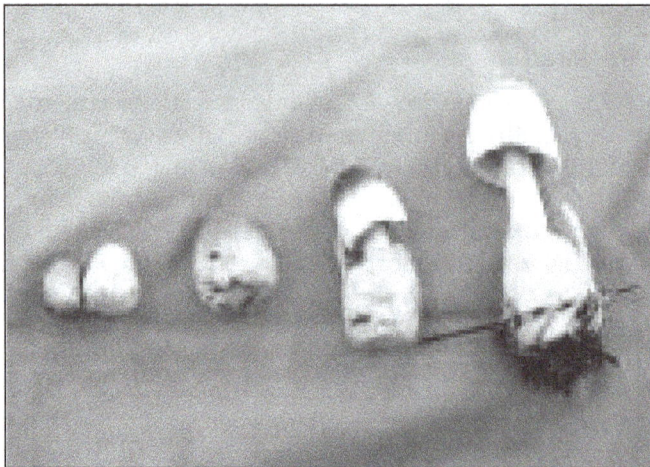

Figure 14.4: Different Growth Stages of Fruiting Bodies of the *Volvariella volvacea* grown on Paddy Straw.

make a single bed which is now pressed to remove the entrapped air and make it compact for effective spawn run.

6. Cropping and Harvest

After spawning, the beds are shifted to the cropping room and covered with polythene or gunny sheets to avoid rapid water loss and maintain required moisture content. The individual beds are watered daily without opening them with a once or twice, depending upon the climatic conditions. The total dry weight of straw/bed is 25 to 32 kg. At least 18 to 22 litre water/bed is sprayed to maintain the moisture level of 65 to 70 per cent. It takes 10 to 15 days for complete mycelial growth under optimum conditions of 30 to 35°C with 85 to 90 per cent RH. Pinheads appear after 4-5 days of spawning. The mushroom is harvested in egg stage (2 inches in diameter), not allowing it to open like an umbrella. First harvesting is done after 9-10 days of spawning and the first flush lasts for 3 days accounting around 75 per cent of the total mushroom yield. The second flush appears after few days and these flush accounts for rest 25 per cent of the total mushroom yield.

7. Post Harvest Management

This mushroom is not recommended to be stored in refrigerated because low temperature storage causes frost injury and deterioration in quality, but can be stored as afresh at a cold temperature of 10 to 15°C for 3 days in polythene bag with perforations. Wooden cases and bamboo baskets also used as packaging for long distance market. This mushroom can be stored more effectively at the button stage than at any other stage. In general the shelf life of this mushroom is very less and mushrooms are sold on the day of harvest.

Chapter 15

Cultivation Technique of Other Edible Mushrooms

Few species of mushroom are also very tasty and gives nutrition and medicinal importance. However the principle and techniques of the production of every species of mushroom is more or less similar, but the cultivation of few species such as *Auricularia* sp., *Flammulina velutipes* and *Tremella fuciformis* etc is a bit complex and tedious and it is cultivated at very low scale. The cultivation technique are being dealt here.

A. Wood Ear Mushroom (*Auricularia* sp.)

This is a popular mushroom in Asia, and well known for the crispy texture. This mushroom is also called Jew's ear, wood ear, tree ear and jelly ear. The fruiting body is distinguished by its distinctly ear-like shape and brown colouration; it grows upon wood, especially elder. The two main species cultivated commercially are: *Auricularia polytricha* and *A. auricula*. It is used as a medicine for the treatment of piles, sore throat and anemia.

Cultivation Technique

This mushroom commonly cultivated on a natural logs of broad-leaf trees or on synthetic medium consisting of sawdust, cotton seed hulls, bran, and other cereal grains.

1. Log Methods

For cultivation on natural logs, members of the oak family (Fagaceae) are preferred, but many other species of both hard and softwoods may be used. The tree are usually 7 to 10 years of age when they are suitable for use. The first step is to punch or drill holes in the logs where spawn will be planted. The holes should be 1.3-2.5 cm deep (the hole should be punched or drilled through the bark and into the hardwood section of the log.) and 3 to 6 cm diameter. Make the holes in lines or a zig-zag pattern all around the log. Simply place the spawn in the hole to the top and pack gently, making sure that the hole is completely full. After complete spawning in the holes, seal the holes with bark, hardwood, plastic or wax plugs. Hardwood plugs are usually better to use because it shrink less and are more likely to stay over the hole for a longer period of time. plastic plugs (available at most spawn outlets) or wax can be used to seal the hole. To use wax as a plug, first melt the wax, then allow it to cool before applying to the log.

The inoculated logs are ready to be incubated for 3 to 4 weeks for the development of mycelium. During incubation period the logs should not be exposed with direct sunlight, wind, or moisture. The logs turned upside down once a month to bring about even mycelial growth. They are also watered whenever necessary. The optimum temperature for the development of the mycelium within the logs is 20 to 28°C. After about 2 months, the logs shifted to the cropping yard which may be open area in a forest, a greenhouse or a hut. Logs are frequently sprayed with water to maintain temperature and moisture. An ideal temperature ranges for cropping yard is 15 to 25°C. Matured mushrooms are ready to harvest in another 28 to 30 days.

2. Sawdust or Plastic Bag Methods

Instead of natural logs, synthetic logs may be used. Synthetic logs are composed of sawdust packed into heat-resistant plastic bags. After proper sterilization, the bags are inoculted with grain or sawdust spawn. This technique is preferable to log technology in regions where logs are not readily available. The method also provides condition where fruiting can easily be controlled by the environment in the mushroom house.

B. Winter Mushroom (*Flammulina velutipes*)

Flammulina velutipes is a popular mushroom in Western Asia, which has fine texture and pleasing appearance. It ranks sixth in terms of total world mushroom production. This mushroom is particularly known for its taste and preventive as well as curative properties for liver diseases and gastroenteric ulcers.

Cultivation Technology

Winter mushroom can be grown during the coldest period of year. This is grown on sawdust (80 per cent) supplemented with rice/wheat bran (20 per cent). Sawdust is wetted thoroughly with water for 16-18 hours. After wetting rice/ wheat bran is added in the sawdust and mixed accurately. Prepared mixture filled in polypropylene bag @ 2 kg/bag. The filled bags are plugged with non- absorbent cotton by inserting a ring on the mouth of the bag and sterilized in autoclaves for 1½ hour at 22 p.s.i. Cooled bag inoculated with spawn @ 4 per cent dry wt. basis. The spawned bag is are incubated at 20-25°C. After 20 to 25 days, when mycelium spreads to 90 per cent of the bag space, the plug of non- absorbent cotton is removed, the neck of the bag is unfolded which made smooth for uniform fruiting. The bags are then placed in the dark at temperature of 10-12°C and 80-85 per cent relative humidity to excite primodial formation. Primordia are formed in 10-14 days after reducing the temperature. At 10-14°C, the fruit bodies grow rapidly, but they are slender, long and of poor quality. For this reason, the growth of fruit bodies is controlled by lowering the temperature to 3-5°C and providing aeration for 1-2 hours daily, which encourage stiff, white and drier fruit bodies. This control is continued for 3-4 days, from the period when the cap's differentiation is observed with the naked eye to the period when the length of the stem reaches 10-12 cm. When the fruit bodies are 14-18 cm long, the fruit bodies are harvested.

A moisture level in the bags is important for fruiting. Primordia are formed in 10-14 days after reducing the temperature to 10-14°C. The initiation of fruit bodies starts in dark but light is necessary for the further development. At 10-14°C, the fruit bodies grow rapidly, but they are slender, long and of poor quality. For this reason, the growth of fruit bodies is controlled by lowering the temperature to 3-5°C and providing aeration for 1-2 hours daily, which encourage stiff, white and drier fruit bodies. This control is continued for 3-4 days, from the period when the cap's differentiation is observed with the naked eye to the period when the length of the stem reaches 10-12 cm. The whole process takes about 50-60 days from spawning to the first harvest. Only two flushes are harvested. About 360-400 g fresh mushrooms can be harvested form a bag of two kg.

C. Silver Ear Mushroom (*Tremella fuciformis*)

Tremella fuciformis commonly known as the white jelly fungus or silver ear mushroom and has been used as a delicacy food in China.

Cultivation Technology

A pure culture of *Tremella fuciformis* can be used for the cultivation on sawdust but mixed culture technique developed in china achieve better yield production. The mixed culture technique involves the use of "helper" mycelium of *Hypoxylon archeri*, commonly associated in nature with decaying wood. *Hypoxylon archeri* increases the ability of *T. fuciformis* to digest the substrate there by increasing mushroom yields. Exploitation of this mycelial association is accomplished through use of dual cultures to make mother spawn.

Substrate used for mushroom production is the same as that used for spawn production. The supplemented substrate is packed into plastic bags (50 cm long and 9 cm diameter) and ends of the bags are tied with cotton string and holes of about 1 cm diameter are made at across the surface for free diffusion of gases and heat generated inside and covered with a breathable fabric. The bags are then autoclaved at 126°C at 20 p.s.i. for 1.30 to 2 hours. Sterilized bag are immediately transfer in the inoculating chamber and allowed to cool down overnight. After proper cooling the bags are inoculated with the mother culture. After about 30 days of vegetative mycelial growth, the hole covers are removed and exposed to conditions favorable for primordial formation. If optimum conditions are maintained in the growing houses, clusters of jelly fungus should be ready for harvest within 12 to 15 days. Yield for each bag of substrate is in the range of 350 to 500 g fresh weight (35 to 50 g dry weight).

Chapter 16

Cultivation Technique of Medicinal Mushrooms

Medicinal Mushroom

Medicinal mushrooms have an established history of use in traditional oriental medicine. Mushrooms have been a subject of modern medical research since the 1960s, where most modern medical studies concern the use of mushroom extracts, rather than whole mushrooms. The significant pharmacological effects and physiological properties of mushrooms are bioregulation (immune enhancement), maintenance of homeostasis and regulation of biorhythm, cure of various diseases and prevention and improvement from life threatening diseases such as cancer, cerebral stroke and heart diseases. Mushrooms are also known to have effective substances for antifungal, anti-inflammatory, antitumor, antiviral, antibacterial, hepatoprotective, antidiabetic, hypolipedemic, antithrombotic and hypotensive activities. (importance of medicinal mushrooms is already mentioned in chapter 2).

Table 16.1: List of Important Species of Medicinal Mushroom

Sl.No.	Spp. of Mushroom	Common Name
1.	Ganoderma lucidum	Reishi mushroom
2.	Lentinula edodus	Shitake mushroom
3.	Cordyceps sinensis	Chinese caterpillar mushroom
4.	Auricularia polytricha	Black ear/wood ear mushroom
5.	Grifola frondosa	Maitake (dancing mushroom)

Cultivation Technique of Shiitake Mushroom (*Lentinula edodes*)

Shiitake mushroom (*Lentinula edodes*) has delicious taste and medicinal attributes and can be considered as king of mushroom and miracle mushroom. It is believed that shiitake cultivation techniques developed in China were introduced to the Japanese by Chinese growers. The shitake mushroom is high prized around the world for its medicinal properties. It is one of the most popular source of protein in many countries.

Figure 16.1: Shiitake Mushroom (*Lentinula edodes*) on Saw Dust.

Taxonomy and Naming

The term *shiitake* is derived from the two Japanese word namely *shii*, the name of the tree *Castanopsis cuspidata* that provides the dead logs on which it is typically cultivated, and *take* meaning "mushroom" The *edodes* is derived from the Latin word for "edible"

Kingdom: Fungi

Phylum: Basidiomycota

Class: Agaricomycetes

Order: Agaricales

Family: Omphalotaceae

Genus: Lentinula

Species: Lentinula edodes

Agro-Climatic Requirements

☆ Best period for cultivation is summer season.

☆ Temperature for cropping: 16-22°C.

☆ Relative humidity of 85-90 per cent.

☆ The major producing states are Odisha, Uttar Pradesh, Punjab, Haryana, Maharashtra, Andhra Pradesh, Tamil Nadu and Karnataka.

Requirement

The require materials are listed below for cultivation technology of shiitake mushroom

Sl.No.	Raw Materials	Sl.No.	Appliances/Tools
1.	Fresh spawn (15 days old)	1.	Autoclave
2	Substrate (saw dust)	2.	Crop room
3.	Supplement (rice/wheat bran)	3.	Water tank
4.	PP bag (60 x 30 cm size)	4.	Cemented floor for drying of straw
5.	Non-absorbent cotton	5.	LPG/Electric stove
6.	Plastic neck	6.	Thermometer and hygrometer
7.	Rubber bands		
8.	Chemicals (Bavistin and formalin, gypsum)		

Cultivation Technique

Shiitake mushroom grows on wheat straw as well as on sawdust substrates, Sawdust substrate gives better fruiting. Sawdust is supplemented with 6 per cent rice/wheat bran and small quantity of gypsum (about 1 per cent dry w/w). The entire technique of Shiitake mushroom cultivation can be done through following steps:

1. Saw dust is soaked for 16-18 hours and wheat bran for three hours. All ingredients are thoroughly mixed.
2. These mixtures are wetted on the composting platform and filled into polypropylene (heat resistant) bags (1.5 to 2 kg), pressed hard to drive out air and a hole.
3. The substrate bags are plugged with non-absorbent cotton with the help of plastic neck on the open side of the bag.
4. The substrate bags are sterilized with autoclave at 121°C at 15 psi for 90 minutes.
5. After cooling the substrates are spawned (using grain spawn) @ 3 per cent (dry weight basis) under aseptic conditions (under laminar flow).
6. After inoculation bags, the fungus mycelium grows in two phases, spawn run and browning.
7. The vegetative growth/spawn run is done at 24±1°C, which takes 30-35 days.
8. The bags after spawn run are opened and maintained in cropping room at 17-19°C, 90 per cent RH and 1000 ppm CO_2 concentration for browning.

9. After browning of 30 days bags are shifted to the cropping room to be maintained at 22±1°C.

10. The complete cycle takes about four months on synthetic logs.

11 The yield potential on synthetic logs is about 80-100 per cent of the dry weight of the substrate.

Cultivation Technique of Reishi Mushroom (*Ganoderma lucidum*)

Ganoderma lucidum, a fungus known by various names like *Reishi*, *Ling Zhi*, and *Mannentake* is one of the pharmacologically and commercially important mushrooms. This mushroom is used as medicine and not as food because it is bitter and corky hard. Market value of this mushroom is basically as herbal medicine and food supplement (nutraceuticals). *Reishi* is reported to possess a plethora of very significant medicinal values; anti-cancer, anti- HIV, anti-heart attack (cholesterol lowering as well as anti-angiogenic), hepato- and nephrotoprotective, hypoglycemic (anti-diabetes), anti-oxidants *etc*. Reishi is considered as panacea in the Chinese and Japanese systems of medicine. It is believed that certain triterpenes and polysaccharides may account for the multiple activities of Reishi.

Figure 16.2: Fruiting Body of *Ganoderma lucidum*.

Taxonomy and Naming

The name *Ganoderma* is derived from the two Greek word: *"ganos"* means *brightness, shining* and *"dermal"* means *skin*. In the year 1881, Petter Adolf Karsten named the genus *Ganoderma*. *Ganoderma lucidum* can be classified according to taxonomy as below:

Kingdom: Fungi

Phylum Basidiomycota

Class: Basidiomycetes

Order: Polyporales

Family: Ganodermataceae

Genus: *Ganoderma*

Species: *Ganoderma lucidum*

Agro-Climatic Requirements

☆ Best period for cultivation is summer season.

☆ Temperature for vegetative growth (spawn run) 20-25°C. and 25-28°C for cropping.

☆ Relative humidity of 85-95 per cent.

☆ The major producing states are Odisha, Uttar Pradesh, Punjab, Haryana, Maharashtra, Andhra Pradesh, Tamil Nadu and Karnataka.

Requirements

The require materials are listed below for cultivation technology of Reishi mushroom

Sl.No.	Raw Materials	Sl.No.	Appliances/Tools
1.	Fresh spawn (15 days old)	1.	Autoclave
2.	Substrate (saw dust)	2.	Crop room
3.	Supplement (rice/wheat bran)	3.	Water tank
4.	Calcium sulphate and calcium carbonate	4.	Cemented floor for mixing supplements
5.	PP bag (60 x 30 cm size)	5.	LPG/Electric stove
6.	Non-absorbent cotton	6.	Thermometer and hygrometer
7.	Plastic neck		
8.	Rubber bands		
9.	Chemicals (bavistin and formalin, gypsum)		

Cultivation Technique

Reishi can be grown by both farmers seasonally under low cost growing rooms preferably polyhouses and by industrialists also under environmentally controlled cropping rooms. This mushroom is intended to be used exclusively as medicine. It has to be grown organically. Seasonal farmers have to put up polycover on the top and sides of the thatched huts and have to maintain hygienic conditions to prevent diseases and pests. Therefore, no toxic chemical should be used for controlling the same. Sawdust is considered as best substrate for cultivation of Reishi mushroom. The entire technique of Reishi mushroom cultivation can be done through following steps:

1. Sawdust supplements are amended with 20 per cent wheat bran and is wetted to a level of 65 per cent moisture.

2 Calcium sulphate (gypsum) and calcium carbonate (chalk powder) are also added to get a pH of 5.5.

3. The mixed substrate (700 g dry wt; 2.1 kg wet) is filled in heat resistant polypropylene bags

4. The substrate bags are plugged with non-absorbent cotton with the help of plastic neck on the open side of the bag.

5. The bags are then autoclaved at 121°C at 22 psi for 2 hours.

6. After cooling the substrate are spawned (using grain spawn) @ 3 per cent (dry wt. basis) under aseptic conditions (under laminar flow).

7. Spawned bags are kept in incubation with 28-35°C in the closed rooms (high carbon dioxide) under dark condition for spawn run.

8. Spawn run (bags white all over), takes about 25 days, thereafter polythene top is cut at the level of the substrate totally exposing the top side and proper condition for fruiting or pinning (temperature 28°C, 1500 ppm CO_2, 800 lux light, 95 per cent RH) is provided.

9. Once the cap is fully formed, which is indicated by yellowing of the cap margin (which is otherwise white). The temperature is lowered to 25°C and RH is further reduced to 60 per cent for cap thickening, reddening and maturation of the fruit bodies.

10. Full maturity is indicated, when the cap is fully reddish brown and spores are shed on the top of the cap.

11. Harvesting is done by the tight plucking, holding the root with one hand and pulling up with another. Scissors and knives can also be used but no residual bud should be left after harvesting.

12. One cycle of the growing takes 10-15 days.

13. After harvesting of the first flush, conditions for pinning are again switched on *i.e.* 28°C, 95 per cent RH, for initiation and completion of the second flush.

Figure 16.3: Different Stages during Growth and Fruiting of *Ganoderma lucidum* on Saw Dust.

14. Depending upon the conditions, 2-3 flushes come out and a total 25 per cent B.E. can be achieved (250 g fresh mushroom from one kg dry substrate).

15. One crop takes about four months.

Chapter 17

Poisonous Mushroom

Poisonous mushroom is also called toadstool but all toadstool are not poisonous; some are harmless, while other are pleasant and edible. There are three types of mushroom poisoning depending upon the principle of toxicity.

1. **Muscarine type** (due to *Inocybe* and related species known as early mushroom poisoning)

2. **Hallucinogenic** (due to *Amanita muscaria*)

3. **Phalloidin type** (due to *Amanita phalloides* and its related species, known as late mushroom poisoning). Mushroom poisoning is also known as mycetism. Fatality due to mushroom consumption is common among poverty ridden families. As such there are no particular features to differentiate between poisonous and non poisonous mushrooms.

Table 17.1: Important Species of Poisonous Mushroom

Sl.No.	Name of Poisonous Mushroom	Common Name
1.	Amanita phalloides	Death cap
2.	A. muscaria	Fly agric
3.	A. pantherina	Panther cap
4.	A. citrina	False death cap
5.	Coprinus atramentarius	Inky cap mushroom

Note: The genus **Amanita** *is one of the most toxic mushroom genera in the world.*

Cautious Point during Eating Mushroom

☆ Uncooked mushroom should not be eaten unless someone is absolutely sure that the variety is edible.

☆ When eating an edible species of mushroom for the first time, someone should take very little amount as a part of toxicity testing.

Table 17.2: Poisonous Mushroom and its Symptoms

Sl.No.	Name of Poisonous Mushroom	Toxin	Symptoms
1.	*Amanita phalloides, A. verna* and *A. virosa*	Cyclopeptides, e.g., Amanitoxins and Phallotoxins	Violent vomiting, diarrhoea and abdominal pain, approximately 12 hours after consumption, which last for days. Remission of symptoms, followed by failure of kidney and liver function, coma and usually death.
2.	*A.muscaria* and *A. pantherina*	Ibotenic Acid, Muscimol, mycetoatropine	Unable to walk or walk with drunken gait, confusion between 30 and 120 minutes of consumption. Alternation between lethargy and hyperactivity. Nausea and vomiting may also occur if too many mushrooms have been consumed. This is followed by a deep sleep with dreams, lasting about two hours
3.	*Coprinus atramentarius, C. insignis* and *C. quadrifidus*	Coprine (Anabuse-like Reaction)	Hot flushes of the face and neck, metallic taste in mouth, tingling sensation in limbs, numbness in hands, palpitations, a throbbing headache, nausea and vomiting. Symptoms begin approximately 30-60 minutes after consumption.
4.	*Psilocybe, Panaeolus* and *Copelandia*	Psilocybin and Psilocin	Variable, affecting the senses. Some of the common symptoms include uncontrollable laughter, hallucinations, euphoria and disembodied experience. Symptoms begin approximately 10-30 minutes after consumption.
5.	*Clitocybe* and *Inocybe*	Muscarine	"PSL" Syndrome (perspiration, salivation, lachrimation) develops rapidly, within 15 to 30 minutes of consumption. Other symptoms include nausea, vomiting and diarrohea, blurred vision and urge to urinate.

Table 17.3: Difference between Edible Mushroom and Poisonous Mushroom

Sl.No.	Edible Mushroom	Poisonous Mushroom
1.	Also known as edible fruiting body	Also known as toad stool fruiting body
2.	Cap is not easily detachable	Easily detachable from the stalk
3.	Fragments of tissues on the cap are absent	Fragments of tissues on the cap
4.	Gills are packed closely together	Gills are free
5.	Stalk are stout with a skirt	Stalk are smooth and narrow upwards
6.	Volva absent	Volva present
7.	Mostly white	Coloured, attractive and beautiful
8.	Mushroom does not change colour when cut or bruise	Mushroom change colour when cut or bruise
9.	Species are *Agaricus, Pluerotus* etc.	Species are *Amanita muscaria, A. phalloides* etc.

1. *Amanita phalloides*

2. *Amanita* sp.

3. *Ganoderna* sp.

4. *Microporous* sp.

Figure 17.1: Few Major Species of Poisonous Mushroom.

☆ Gathering and eating wild button mushroom must be avoided.

☆ Eating of over-matured, insect-pest infested, wilted wild mushroom must be avoided.

Most of the toxic varieties of mushrooms contain more than one of the following classes of toxins: monomethylhydrazines, cyclopeptides, ibotenic acid, psilocybin, coprine, muscarine, and unknown toxins. First an effort should be made to identify the particular mushroom responsible for the poisoning preferably with the help of a mushroom expert. The clinical syndromes of the patients generally depend on the ingestion of predominant toxin, which have been summarized in the Table 17.2.

Chapter 18

Post Harvest Management of Mushroom

Like all fruits and vegetables, mushrooms are perishable and have very short self-life after harvest, they often change in ways include wilting, ripening, browning, liquefaction, loss of moisture, and loss of texture, aroma and flavour that make them unacceptable for human consumption, hence processing is recommended to increase their shelf-life. Their shelf life can be enhanced by processing them. Processed mushrooms are packed in special quality polythene bags or canned. Initially, fresh mushrooms are washed in cold water and then blanched in boiling water for about 3-4 minutes. Then they are dehydrated in a drier and packed. It is advisable to pre-treat fresh mushrooms in a solution containing brine to prevent discolouration. Packing is very critical as formation of moisture contamination in mushrooms very quickly. Yield of dried mushroom depends upon many factors like moisture content in fresh mushrooms, type of dryer, process employed and moisture content required in the finished product *etc.* Hence, average yield is taken at 25 per cent. A brine of 2 per cent salt and 0.2 per cent citric acid are used for packing. The cans are exhausted at 19°C for 7-8 minutes, sealed and processed under pressure for 20-25 minutes.

Processing

☆ Mushrooms are washed in cold water to remove soil particles.

☆ Blanching is done in stainless steel kettles filled with a boiling solution of 0.1 per cent citric acid and 1 per cent common salt for 5-6 minutes at 95-100°C to inhibit polyphenol oxidase enzymes activity and to inactivate microorganisms.

☆ Dehydration is done to bring moisture level below 10- 12 per cent at 55-60°C. Doing this insects and microbes are killed.

☆ Controlled atmosphere packaging (CAP) and modified atmosphere packaging (MAP) are most preferable practice for mushroom packaging.

☆ For short distances, the polypacks of mushrooms can be stacked in small wooden cases or boxes with sufficient crushed ice in polypacks, while for long distance, transport of large quantities in refrigerated trucks.

☆ Labelling must be done either manually or mechanically

☆ Storage of mushrooms at 0 to 2°C with 95 per cent RH might keep mushroom preserved for 4 to 5 days

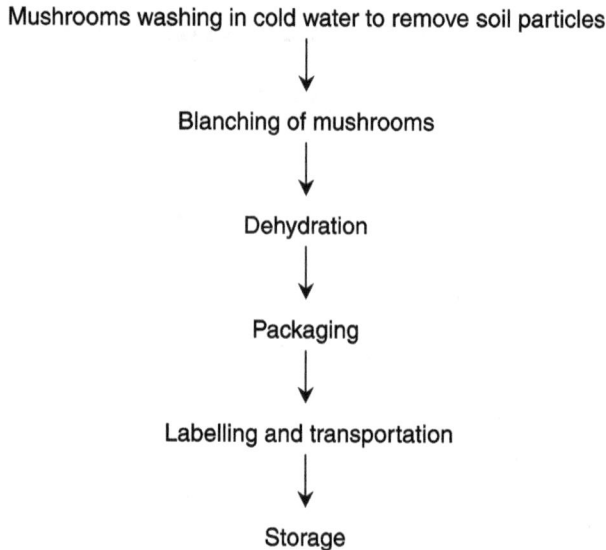

Mushrooms washing in cold water to remove soil particles

↓

Blanching of mushrooms

↓

Dehydration

↓

Packaging

↓

Labelling and transportation

↓

Storage

Figure 18.1: Flow Chart of Processing of Fresh Mushroom.

Canning

Generally canning is used for preservation of mushroom, particularly *Agaricus* mushroom. Canning is the technique by which the mushrooms can be stored for longer periods up to a year and most of the international trade in mushrooms adopts this procedure.

☆ Well graded fresh white colour mushrooms must be taken.

☆ Whole mushrooms must be washed 3-4 times in cold running water to remove adhering substances.

☆ Blanching in stainless steel kettles filled with a boiling solution of 0.1 per cent citric acid and 1 per cent common salt for 5-6 minutes at 95-100°C inhibit polyphenol oxidase enzymes activity and to inactivate microorganisms.

☆ After blanching, mushroom are filled in sterilized tin cans (A-2½ and A-1 tall can sizes containing approximately 440 and 220 g drained mushroom weight, respectively).

☆ After filling, the cans are exhausted by passing them in exhaust box for 10-15 minutes, so that temperature in the centre of cans reaches up to 85°C.

☆ Then the cans are sealed hermetically with double seamer and kept in upside down position.

☆ After exhausting of cans, sterilization of cans up to 118°C is done to prevent the spoilage by microorganisms during storage.

☆ The cans are cooled immediately in a cold-water tank after sterilization process to stop the over-cooking and to prevent stack burning.

☆ Thereafter the clean and dry cans are labeled manually or mechanically and packed in strong wooden crates or corrugated cardboard cartons.

☆ The cans are stored in cool and dry place before dispatching to market.

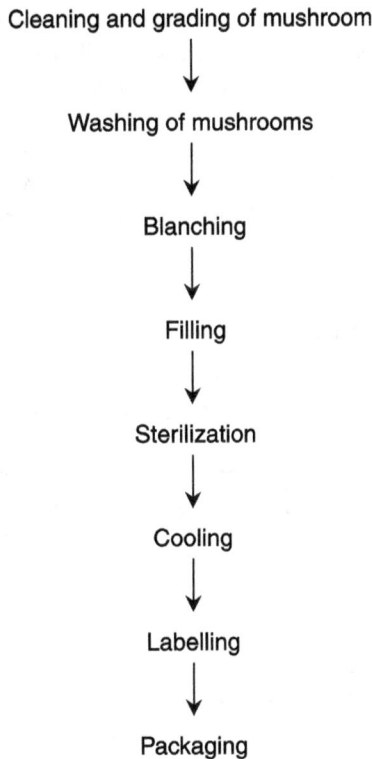

Cleaning and grading of mushroom
↓
Washing of mushrooms
↓
Blanching
↓
Filling
↓
Sterilization
↓
Cooling
↓
Labelling
↓
Packaging

Figure 18.2: Flow Chart of Canning.

Drying

Drying is the process of exclusion of moisture from the product to such a that the microbial and biochemical activities are checked due to reduced water activity, which makes the products suitable for safe storage and protection against the attack by microorganisms during the storage. Mushroom contains about 90 per cent moisture at the time of harvesting and are dried to a moisture level down below 10-

12 per cent. There are numerous drying methods such as sun-drying, thermal power draying, freeze drying, cabinet air drying, osmo-air drying, fluidized-bed drying and microwave drying. Sun-drying is very oldest, cheapest and very easy method among various drying methods. Under this method, mushrooms are cut off at the basal part of the stalk and spread over the trays or sheets and kept in open under the sun with 25-30°C temperature, with less than 50 per cent relative humidity and high wind velocity. Usually about 2-4 days under continues daily sunshine is adequate for sun draying. Sun dried mushroom product contains more than 10-12 per cent moisture and should therefore be oven dried at 55-60°C for 4-6 hours to further reduce the moisture to 7-8 per cent to avoid whichever spoilage during storage. Under freeze drying; removal of water from a substance by sublimation from the frozen state to the vapour state is known as freeze-drying. The product is frozen at -22°C for one minute. The frozen mushrooms are dried to moisture content of 3 per cent in a freeze drier and packed under vacuum. Mushroom should not be dried under high temperature because it loose texture, flavor, color along with reduced rehydrability. Dried mushrooms can be easily powdered and used in soups, bakery products *etc*. Therefore to enhance the post harvest life, dried mushroom should be put into polythene bag, sealed and kept in a dry, cool, and dark place. For long term storage, the mushroom should be packed in cartons or wooden boxes and kept at 2 to 5°C in a low temperature. Most of the mushrooms except the button mushroom and milky mushroom have been conventionally dried for long-term storage *e.g. Pleurotus, Volvariella, Lentinus, Auricularia etc.*

Packing and Packaging

The mushrooms are packed for transporting them to the market. While a good package sells a product, a mediocre package can interfere with sale of an otherwise excellent product. For the local markets in India, mushrooms are packed in retail packs of 200 g or 400 g in simple polythene packs of less than 100 gauge thickness. Large quantity packing of mushrooms is done using polythene or pulp-board punnets, which will withstand long distance transport. Plastic punnets 130 x130 x 72 0.40, Cardboard chips 305 x 125 x 118 1.82, Plastic tray 330 x 280 x 145 2.30, and Expanded polystyrene 400 x 333 x 167 4.56. These punnets are over-wrapped with differentially permeable PVC or polyaccetate films, which help in creating modified atmosphere in punnets with 10 per cent CO_2 and 2 per cent O_2 and mushrooms maintain their fresh look for 3 days at 18°C. Mushrooms should be packed in cartons with a perforated over-wrap of polyethylene film to reduce moisture loss. It is important to avoid water condensation inside packages.

In the recent years controlled atmosphere packaging (CAP) and modified atmosphere packaging (MAP) are catching up fast for all types of fruits and vegetables. These packaging techniques have to be effectively used for packing mushrooms to have improved shelf life. If simple polythene bags are used, it is important to make desired number of holes for proper humidity control.

Storage

The storage condition is very important for maintaining the quality of fresh mushroom. The best result for storing fresh mushroom in a cool chamber. Depending on the species, the shelf life of mushroom may vary from 1 day to 2 weeks. The shelf life of fresh mushroom may be extended by refrigeration at 1-2°C. Refrigerated storage is however, not always effective, particularly with some tropical or subtropical mushroom which suffer from chilling injury. Temperature, RH, CO_2 is essential factor under storage. High RH is essential to prevent desiccation and loss of glossiness.

In the recent years controlled atmosphere storage is most applicable to increase prolonged shelf life of mushroom. In this method, the oxygen and carbon dioxide concentrations are altered inside the package and respiration rate gets altered. Controlled atmospheric package reduces brown discolouration (enzymatic browning) and extend shelf life. The calcium chloride added to the irrigation water improves the yield and colour of the canned mushroom product. Mushrooms exposed to lower airflow rates in the growing room remain whiter than those exposed to higher airflow rates. Addition of proprietary form of stabilized chlorine dioxide (oxigne, 200 ppm) to pre-harvest watering applications on mushrooms had lower incidence of blotch and mushrooms retained whiter colour during storage. Commercially available food grade moisture absorbers such as montmorillonite clay and silica gel can be used to extend the shelf life of mushrooms in packs. Mushrooms treated with honey (0.5 and 1 per cent) for 18 hours, air dried to remove surface moisture, packed in 100 gauge polythene pouches with 0.5 per cent venting area, increased the shelf life by more than a week over control at 3-5°C and 2-3 days at ambient temperature.

Shelf life of button mushroom is not very excellent however, we can store them for one or two days by keeping them in damp cloth or paper towel in the refrigerator. In general the shelf life of paddy straw mushroom is also very less and mushrooms are sold on the day of harvest.

Transport

The effect of pre-cooling and packing will be partially negated if the product is later stored and transported in a warm environment. Mushrooms, therefore, need refrigerated transport. To keep the mushrooms cool during short distances transport, the polypacks of mushrooms can be stacked in small wooden cases or boxes with sufficient crushed ice in polypacks (over-wrapped in paper), while transport of large quantities in refrigerated trucks is essential for long distance. For long distance, transport of large quantities in refrigerated trucks is essential though it is costlier. Mushroom may also be packed in bamboo basket and transported by airfreight. For transporting oyster mushroom the fruit bodies are stacked in trays or baskets covered with thin polythene sheet with perforation. Shelf life of button mushroom is not very excellent and require cool chain transport for delivery to the market place as button mushroom browns quickly. but specialty mushrooms (except only *Volvariella*) can be transported in ambient conditions with modified packaging, without any damage to the quality of the mushrooms. *Volvariella* mushroom has very less shelf life and are sold on the day of harvest. Use of bamboo baskets with an ventilation channel at the center and dry ice wrapped in paper placed above the mushrooms, is in best way for transportation of *Volvariella* sp. Milky mushroom has a superior shelf life, and can be transported to short distances for marketing without damage to its quality.

Chapter 19

Mushroom Cooking and its Value Addition

Mushroom Cooking

From ancient times our ancestors had been collecting and consuming wild mushrooms. Even today in few regions of our country people collect wild mushrooms. Luckily now many of mushrooms accepted as edible purpose and cultivated throughout the world such as button, oyster, milky, shiitake, paddy straw are common household name. Today mushrooms are more common in menu of parties and restaurants and other celebrations then in household kitchens. A common man or a housewife is still hesitant to cook mushrooms due to less exposure about its cooking methods. Creative cooking with mushroom has a lot of potential in our kitchens. We need to spread this among the masses that mushrooms are not only tasty but have lot of nutritive and medicinal values. They blend well with other vegetables and products. In few regions of the country, it is still associated with non-vegetarian food. To overcome with this myth we really need to popularize mushroom recipes and its cooking methods.

Use Mushrooms in Cooking

In India, many methods by which the mushroom can be used for cooking purpose such as pressure cooking, toss or soute, microwave, grill and boiling or blanching. Pressure cooking is the most common cooking method of Indian kitchen Table 19.1). For cooking mushroom in pressure cooker only one pressure is enough as mushroom has the quality not to dissolve with liquids or gravy but retain same shape. Over cooking can make mushroom more chewy to eat. Boiled/blanching mushrooms are mainly used for making pickles or sandwich fillings. Microwaving is a good idea for oil free cooking. Few mushrooms like button and oyster are ued as raw with salad and sandwitch. This mushroom can be sliced, diced, cut into halves or the whole pieces of small mushrooms can be used as raw.

Mushroom needs to be washed thoroughly before cooking. They can be used as whole in few of the recipes like mushroom munchurian, cut into halves or quarters, can be sliced into thick or thin slices and can be used in soups and salads. Mushrooms can be coarsly or finally chopped for sandwiches, pakodas and koftas. In few of the recipes caps of the big sized mushrooms can be used such as stuffed mushrooms in which we can use different kinds of fillings to stuff mushrooms and bake them. Oyster and shiitake mushrooms can be minced in food processor for use in koftas or pakodas. Do not wash the mushrooms if you need to store them for one or two days, preferably keep them in a paper bag and store in a refrigerator.

Table 19.1: Mushroom Cooking Principle

Mushroom	Cooking Quality	Cooking Principle
Button mushroom	It has excellent culinary qualities	Fresh button mushrooms are being used as cooking but are also available in the market as canned mushrooms. This mushroom can be used as whole or cut into pieces for cooking.
Oyster mushroom	Flavour and velvety texture of this mushroom is liked by most of the people.	This mushroom can be eaten as cooking and raw as salad but it tastes better after cooking by any method. Wash oyster mushroom in hot water and drain extra water properly by squeezing. After washing it can be chopped into small pieces with knife or hand tearing in the direction of gills. Oyster can also be minced in the food processor for making few recipes like koftas or munchurian.
Paddy straw mushroom	This is very tastiest and delicious mushroom	Wash this mushroom under running water for 2 minutes as this is a very delicate mushroom. Cut into small pieces and use it in the dishes of main coarse. This mushroom may not be consumed as raw.
Milky mushroom	This mushroom has a very strong aroma and meaty texture.	Boil this mushroom in salty water for 10 minutes to remove its pungency. After boiling, thoroughly drain the water by squeezing. This mushroom takes more cooking time as it is little harder to cook. This mushroom may not be consumed raw.
Shiitake mushroom	This mushroom is used as medicine and not as very much used as food because it has bitter and leathery texture.	In India, this mushroom is not very much popular for eating purpose because due to its bitter flavour and leathery texture. This mushroom can also be cooked by any method. Boil this mushroom for 2-3 minutes and drain water. After this it can be striped into small pieces. Shiitake mushroom tastes very good in soups. This can also be blended with other vegetable such as capsicum, peas and carrots. This mushroom is the second most popular mushroom in the world having medicinal properties.

Value Addition in Mushroom

Fresh mushrooms have very limited life and hence processing is recommended to enhance their shelf-life. Due to presence of more than 90 per cent moisture content, mushrooms are highly perishable and start deteriorating immediately after harvest. They develop brown colour on the surface of the cap due the enzymatic action of phenol oxidase, this results in shorter shelf-life. Loss of texture, development of off flavour and discoloration result in poor marketable quality and restricts trade of fresh mushrooms.

In view of their high perishable nature, the fresh mushrooms have to be processed to extend their shelf life for off season use. This can be achieved by adopting appropriate post-harvest technology to process surplus mushrooms into novel value added products. The value-added products are the need of the hour for the mushroom growers not only to reduce the losses but also to enhance the income by value-addition and boost the consumption of mushroom. The possible value-added products can be developed either by converting freshly harvested mushrooms into ketch-up, murabba, candy, chips and pickles or by dehydrating freshly harvesting mushrooms into dehydrated form and then making soup powder, biscuit, nuggets and RTE.

Mushroom Ketch-up

☆ Freshly harvested button mushrooms are washed in 0.05 per cent potassium meta bisulphite (KMS) solution, sliced and cooked in 50 per cent of water for 20 minutes.

☆ Making mushroom paste is prepared using a mixer grinder with 0.2 per cent arrarote, 1.5 per cent acetic acid and other ingredients and cooked to bring its TSS to 35° Brix.

☆ Then the ketch-up is filled in the sterilized jars.

Followings are the ingredients that are used for preparation of ketch-up:

Sl.No.	Ingredients	Quantity
1.	Salt	10 per cent
2.	Sugar	25 per cent
3.	Acetic acid,	1.5 per cent
4.	Sodium benzoate	0.065 per cent
5.	Onion	10 per cent
6.	Garlic	0.5 per cent
7.	Ginger	3.0 per cent
8.	Red chilli powder	1.0 per cent
9.	Ajinomoto	0.2 per cent
10.	Arrarote	0.2 per cent
11.	Cumin	1.0 per cent
12.	Black pepper	0.1 per cent

Mushroom Candy

☆ Freshly button mushrooms are graded, washed and halved into two pieces.

☆ Halved pieces are blanched for 5 minutes in 0.05 per cent of KMS solution.

☆ After draining for half an hour, they are treated with sugar. Sugar treatment is given at the rate of 1.5 kg sugar per kg of blanched mushroom. Initially sugar has to be divided into three equal parts. On the 1st day, blanched mushrooms are covered with one part of sugar and kept it for 24 hours. Next day, the same mushrooms are covered with 2nd part of sugar and again kept for overnight and on the third day mushrooms are removed from the sugar syrup. This sugar syrup is boiled with 3rd part of sugar and 0.1 per cent of citric acid to bring its concentration up to 70° Brix.

☆ Blanched mushrooms are mixed with this syrup and again the contents are boiled for 5 minutes to bring its concentration up to 72° Brix.

☆ After cooling, the mushrooms are removed from the syrup and drained for half an hour.

☆ The drained mushrooms are placed on the sorting tables to separate only defected and unwanted pieces and are subjected to drying in a cabinet drier at about 60°C for about 10 hours.

☆ As soon as they become crispy, all mushrooms are taken out and packed in polypropylene bags.

☆ The candy can be stored up to 8 months with excellent acceptability and good taste.

Mushroom Murabba

In preparation of 1kg mushroom murabba, 1.250 kg of sugar is required and cooking is continued till a concentration of at least 68 per cent of soluble solid is reached.

☆ Freshly button mushrooms are graded, washed and pricked.

☆ Blanched in 0.05 per cent potassium meta bisulphite (KMS) for 10 minutes.

☆ Treated with 40 per cent of its weight of sugar daily for 3 days.

☆ Then, mushrooms are taken out from the syrup and 0.1 per cent citric acid and remaining 40 per cent of sugar is mixed in the syrup.

☆ After making its concentration to 65° Brix, mushrooms are added in the syrup and the good quality murabba is prepared.

Mushroom Chips

☆ Freshly harvested button mushrooms are graded, washed and sliced. This is blanched in 2 per cent brine solution. And dipped overnight in a solution of 0.1 per cent of citric acid +1.5 per cent of NaCl + 0.3 per cent of chilli powder.

☆ After draining off the solution, the mushrooms are subjected to drying in cabinet dryer at 60°C for 8 hours.

☆ Then it is fried using the refined oil, to get good quality chips. Garam masala and other spices can be spread over the chips to enhance the taste. After spice mixing, the chips are packed in polypropylene packets and sealed after proper labelling.

Mushroom Nuggets

Mushroom is converted into mushroom powder, mixed with the urd dhal powder and a paste made adding water. Ingredients are added to the prepared paste and round balls of 2 to 4 cm diameter are made out of the paste. Balls are spread over a tray and are sun dried. Now the mushroom nuggets are ready.

Ingredients used in the preparation of mushroom nuggets:

Sl.No.	Ingredients	Quantity
1.	Urd dhal powder	80 per cent
2.	Mushroom powder	10 per cent
3.	Salt	02 per cent
4.	Red chilli powder	01 per cent
5.	Sodium bicarbonate	0.01 per cent
6.	Water	07 per cent

Ready-to-Eat Mushroom Curry (RTE)

It is generally prepared from freshly harvested mushrooms. But it can also be prepared from dried button mushroom slices after its rehydration adding the following ingredients:

Sl.No.	Ingredients	Quantity
1.	Onion	510g
2.	Green chilli	250g
3.	Garlic	250g
4.	Ginger	200g
5.	Salt	160g
6.	Red chilli powder	150g

Mushroom Biscuit

Mushroom biscuit is prepared from mushroom powder by mixing it with following ingredients.

Sl.No.	Ingredients	Quantity
1.	Maida	100g
2.	Fat	45g
3.	Sugar	30g
4.	Milk powder	1.5g
5.	Glucose	1.5g
6.	Baking powder	0.6g
7.	Salt	0.6g
8.	Ammonium bicarbonate	0.3g

Chapter 20

Major Diseases, Pests, Disorders of Mushroom and their Management

Mushrooms, like any other cultivated crop, is subject to attack by pathogens. Cultivated mushrooms are attacked by a number of fungal and bacterial diseases that may cause significant production losses. These occurrences are due to incorrect handling, environmental conditions *etc.* to which the mushroom cultivation is generally carried out. Some important encountered diseases of mushroom and their management have been summarized in the Table 20.1.

General Instruction for the Management of Mushroom Diseases

1. Lime should be sown on the shelves before putting a new bag of mushrooms. This action is intended to kill the bacteria found in the vicinity of shelves and the mushroom house.
2. Placed chalk powder around the infected bag of mushrooms to avoid infection of other fungi.
3. Mushroom house need clean, orderly and free from fungus, bacteria, insect pest and animal. Thus mushroom house should be treated with care.
4. Damaged, infected, problematic and expired mushroom bag to be taken out from the mushroom house.
5. New bag brought into the mushroom house to replace the expired or old mushroom bag.

Table 20.1: Major Diseases and Disorders of Mushroom and its Management

Name of Disease	Causal Organism	Symptoms	Management
		A. Fungal diseases	
1. Dry bubble	*Verticillium fungicola*	Whitish mycelial growth initially appears on the casing surface which has the tendency to turn greyish yellow with age. In mature stage, the mushroom fruit are deformed .	1. Sterilized casing soil and spent compost should be used. 2. Affected patch must be sprayed with 2 per cent formalin. 3. Spraying with dithane Z-78 (0.15 per cent) at weekly interval after 7-8 days of casing.
2. Wet bubble	*Mycogyne perniciosa*	Symptoms in the form of white mouldy growth on the mushrooms, leading to their putrefaction (giving foul odour) with golden brown liquid exudates are also observed. These water drops later change into amber colour.	1. Use of only sterilized casing soil with 1 per cent formalin before 2-3 days of its application followed by immediate spray of carbendazim or benomyl 0.1 per cent after casing is recommended. 2. Heating and fumigation of mushroom house is a good laboratory practice. 3. Spraying of dithane Z-78 (0.15 per cent) at weekly interval is also better.
3. Green mould	*Trichoderma* sp.	It is the most common disease in oyster mushroom where green coloured patches appear on mushroom bags and spawn bags are observed on cubes. It most generally comes in the air or from human handling.	1. Dipping a cotton swab in formalin solution (2 per cent) and scrapping off the affected area might control the disease. 2. If cube attacks more than half, then the entire cube should be discarded. 3. Contaminated cube must be burnt or buried in a place far from the cropping room to avoid re-infection. 4. Use of fresh and uncontaminated spawn is recommended.
4. Cob web	*Hypomyces rosellus*	First appearance of small circular patches of grayish white mycelium on the casing surface. As the disease progresses, a fluffy white mycelium grows over the mushrooms which look like a cotton balls. Eventually they turn brown, begin to rot and die-off	1. Use of sterilized casing soil with 1 per cent formalin before 2-3 days of its application is the best practice to control this disease. 2. Spraying with dithane Z-78 (0.15 per cent), carbendazim (0.2 per cent) between the flushes might be successful control measure. 3. Maintain humidity in the mushroom house.

Contd...

Table 20.1–Contd...

Name of Disease	Causal Organism	Symptoms	Management
5. Yellow mould	*Myceliophthoral utea*	Brownish yellow corky layer of mycelium (stroma) with a white fluffy edge generally observed at the junction of compost and casing.	1. Yellow moulds generally observed where the compost has more than 70 per cent moisture and more than 20°C temperature in the crop room. Moisture and temperature level must be controlled.
6. White plaster mould	*Scopulariopsis fumicola*	Dense white patches of mycelium on compost and casing soil can be seen, giving flour like appearance.	1. If the compost retains smell of ammonia and has pH more than 8.0, white plaster moulds become common. Ammonia free compost must be used.
B. Bacterial Diseases			
1. Yellow blotch	*Pseudomonas tolassi*	It is serious disease of white button mushroom. Formation of pale yellow lesion on mushroom tissues which later become a golden yellow or rich chocolate brown. The disease is characterized by primordia, with yellow droplets on their surface, which become stunted, yellow to orange, and deformed as they mature.	1. Manipulation of relative humidity, temperature, air velocity and air movement are significant in the control of this disease.
2. Mummy Disease	*Pseudomonas* sp.	The stalk is bent and the cap tilted. Often there is a dense growth of mycelium around the base of the stalk on the surface of the casing layer. Mushrooms often fail to mature and remain in the "button" stage with unopened veil.	1. Strict hygienic condition in mushroom house 2. Disinfecting the casing layer can help to reduce the infection.
C. Viral Diseases			
1. La France	Virus	A specific musty smell in a growing room. Mushroom fruits appear in dense clusters, maturing too early.	1. Hygienic measures should be followed strictly. Newly virus free mushroom fungus culture should be used.

Contd...

Table 20.1–Contd...

Name of Disease	Causal Organism	Symptoms	Management
D. Weed			
1. Ink caps	Coprinus sp.	It is a weed of mushroom that develops on the cubes before cropping begins. It subsequently disintegrates into a black sliming mass at maturity.	1. Before filling the trays the compost should be free from ammonia. 2. Avoid excessive watering. 3. Properly pasteurized compost and casing soil should be used. 4. Physical removal of Coprinus from the cube.
E. Physiological Disorders			
1. Stroma and sectors	It is due to genetic characters or mishandling of spawn	It is the aggregation of mushroom mycelium on the surface of the spawned compost or the casing. It is like a extra-white, extra-dense of fluffy and always different from the normal spawn.	1. Care should be taken during handling of mushroom spawn. 2. Do not put the chemical, detergent and petroleum based product near the spawn.
2. Flock	This disorder is due to improper environment conditions under the mushroom house.	It is the type of malformation appear in fruit cap and tissues of the gills. The cap opens prematurely and affected gills are rudimentary.	1. Maintain proper environmental condition in the mushroom house. 2. Do not put the chemicals, oil based paint fumes, detergent and petroleum based product in the mushroom hut.
3. Hollow core and brown pith	It seems due to excessive watering	A circular gap seen in the centre of the stem. This gap may be extend the length of the stipe or it may be shorter.	1. Hygienic measures should be followed strictly 2. Avoid excessive watering
4. Brown discolouration	It is due to high temperature, watering with high pressure and incorrect use of formalin	Browning discolouration can be seen on small pin heads or immature mushroom fruit.	1. Maintain proper environmental condition in the mushroom house. 2. Avoid watering with high pressure. 3. Care should be taken during the use of formalin.

Contd...

Table 20.1–Contd...

Name of Disease	Causal Organism	Symptoms	Management
5. Scales or crocodiles	It is due to uncontrolled climatic condition in the mushroom house. Strong vapours of formaldehyde may also cause this problem.	Scales like structure arise on the surface of mushroom fruit tissues which fail to develop mushroom fruit cap. As the mushroom continues to grow, the skin bursts and so called crocodile skin is formed.	1. Maintain proper environmental condition in the mushroom house. 2. Care should be taken during the use of formalin. 3. Avoid excess vapours from the formaldehyde.
6. Abnormal fruit bodies	Presence of high-level of carbon dioxide.	This abnormal fruit bodies always different from the normal.	1. Avoid excess CO_2 in the mushroom house.

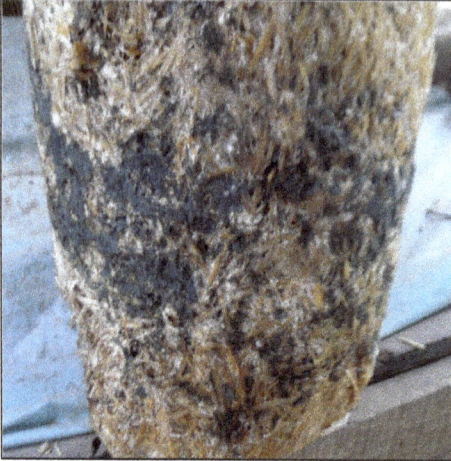

1. Green mould (*Trichoderma* sp.)
on mushroom bag

2. Green mould (*Trichoderma* sp.)
in spawn bag

3. Dry Bubble (*Verticillium fungicola*)

4. Wet bubble (*Mycogyne perniciosa*)

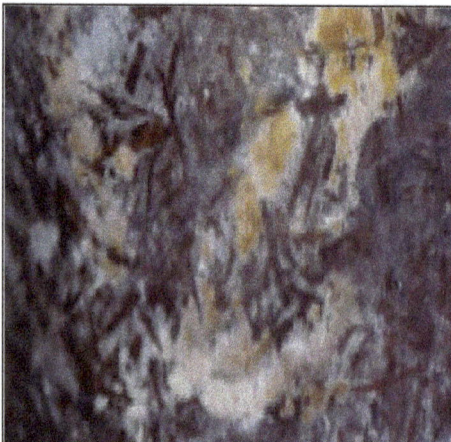

5. White plaster moulds
(*Scopulariopsis fumicola*)

6. Bacterial yellow blotch
(*Pseudomonas tolassi*)

Figure 20.1: Diseases of Mushroom.

Table 20.2: Major Insect and Mites of Mushroom and its Management

Name of Insects	Scientific Name	Damaging Characters	Management
A. Insects–Flies			
Sciarid fly	Lycoriella mali	A small black insect about 0.25 inch long, with long antennae and gray wings folded over the back. Sciarid larvae attack compost, spawn, mycelia, pins, and mushroom stems and caps.	1. Nylon or wire net (not less than 35 mesh) should be placed at window to prevent the entry of flies into the crop rooms.
			2. If flies present in crop room before mixing thoroughly fipronil 0.3 per cent GR @ 100 g in the 100 kg ready casing soil.
Phorid fly	Megaselia halterata	These are small, light to dark brown colour, 1.9–2.0 mm long, with a humpback appearance and very small antennae. They are stockier than sciarids and are very active, running and hopping erratically. These flies feed on mycelia, and restrict spawn run. The infested mushrooms turn brown and leathery with rotting tissues.	3. During composting, mix thoroughly fipronil 0.3 per cent GR @ 50 g/qtl. of wheat straw during 7th turning.
			4. Spray the affected patch with 0.2 per cent formalin.
Cecid flies	Mycophila speyeri	Cecids feed on the mushroom stems or gills, reducing marketable yield.	
Springtails	Seira iricolor	Adults are observed to be ground colour with light violet along side of the body without forming a definite pattern. The adults measured 2.9 mm longs. Adults and nymphs feed on mycelium by scraping from the spawn grains and cutting the mycelial strands.	1. The surrounding and inside of crop room must be kept neat and clean.
			2. Crop beds should be raised-off from the floor.
			3. Compost and casing soil should be properly sterilized
			4. Area surrounding mushroom house should be sprayed with fipronil 5 per cent SC @ 0.1 ml/lit. of water.
B. Slugs and Snails			
		These pests chew up portion of the mushroom which may later get infected with bacteria and affect the quality of the crop.	1. Remove the pests from the cubes and kill.
			2. Maintain hygienic conditions.

Contd...

Table 20.2–Contd...

Name of Insects	Scientific Name	Damaging Characters	Management
		C. Mites	
1. Sporophagous mites	*Tryophagus putrescentiae*	It feeds on mycelium and damage sporophores by causing shrunk caps and brown rusted spot on buttons.	1. Cleaning of the mushroom house and disposal of all organic debris must be a common practice.
			2. Mites are usually introduced into the compost by flies during migratory stage.
2. Mycophagus mites (Red pepper mites)	*Pygmephorus* sp.	These are tiny, yellowish-brown in colour often swarm in vast numbers on the surface of the casing and mushroom fruit. They cling to the body of mushroom flies and are spread from infested to uninfested mushroom farms.	3. Efficient composting and peak-heating of fresh compost can avoid mite infestation.
			4. Disinfest the mushroom house by spraying floor, wall and sprayed with sulphur dust @ 0.2 g/lit. of water.
			5. If mites are present in the compost, spray with sulphur dust @ 2 g/lit. of water.
		D. Nematode	
	Aphelenchoides composticola, Ditylenchus myceliophagus and *Rhabditis lambdiensis*	Nematode attack compost become devoid of spawn smells fouls and sunken areas appear on the mushroom bed surface. Nematodes feed the contents of the hyphea, destroying them and turning the compost sudden. They also eat the mushroom and turn them brown/watery.	1. Soil should be sterilized by steam (70-75°C for 6 hours) or formaldehyde 40 per cent (5 per cent solution).
			2. Adding and mixing of nemagon @ 40 ml/30 kg of wheat straw compost.
			3. Mixing of thionazin @ 80 ppm in composting material must be done at the time of spawning.
		E. Rodents	
		The attack by rodents is found mostly in low cost mushroom house (mud house). They eat the grain spawn and make holes inside the cubes.	1. Application of rat trap or sticky trap.
			2. Burrow of rats should be close down with glass pieces and plaster.
		F. Snake	
		They make hole in the mushroom bag for habitat and destroy mycelium spread in the substrate.	1. Broadcasting of fipronil 0.3 per cent GR as borderline outside of the hut.

Major Insect, Mites and Nematodes of Mushroom and their Management

Cultivated mushrooms can be infested by phorid flies, sciarid flies, cecids, springtails, mites and nematodes that may cause serious problems in mushroom cultivation and management of those pests is a key factor in successful mushroom production. These pests damage the mushroom right from spawning to harvesting. Mushrooms being an indoor crop, provide a suitable habitat for the insect-pests where pests remain protected from the vagaries of the weather. Under Indian conditions, most of the growers cultivate mushrooms seasonally and hardly pay any attention to hygiene and sanitation aspects. These seasonal farms lack air handling units and fresh air is introduced by keeping the doors of cropping rooms open daily for 2-3 hours. As a result, a large number of mushroom pests gain easy access into these farms. Therefore, measures should be taken to prevent the entry of insect pests, mite and nematode into the cropping rooms. However, chemical method of pest control should be taken as last step. Some important insect and pest of mushroom and their management have been concise in the Table 20.2.

Chapter 21

Ethics and Economics in Mushroom Cultivation

Mushroom cultivation is one of the agricultural activity in which rural youth can play a vital role without sacrificing their household responsibilities. It can provide self-employment and earning additional income for both the semi-urban and rural areas, especially suitable for small, marginal poor farm household, farmwomen, landless labourers, rural unemployed youth and even retired or in-service personnel in order to raise their socio-economic status. Apart from the socio-economic status of the mushroom growers, it not only solve the problems related to employment of both literate and illiterate person (specially women) *vis-a-vis* considered as an alternative source of income to uplift the living standards of poor people and also to add high quality protein in their daily diets to eradicate malnutrition problems. There is less skilled labour and capital required during the initial scale. Its relevance in the present scenario can be examined from the view that the agricultural land will decrease due to industrialization, population pressure and conversion into wasteland due to intensive agriculture. The mushroom can be cultivated in air since, the entire cultivation is an indoor affair, hence, it is highly women friendly in nature and will provide them with more opportunities for cultural, societal, and technical education in improving the quality of family and community life by income generation.

Entrepreneurship

Agriculture is the livelihood of majority of the rural population of India, with over 80 per cent of the cultivators as small and marginal farmers. High labor-land ratio and alarming rate of population growth may pose a threat to our food security in the very near future. In addition to their role in agricultural production, women are gainfully employed in agri-based allied activities like dairying, animal husbandry, poultry, goatery, rabbitry, beekeeping, mushroom cultivation, floriculture, horticulture, fruit preservation, post-harvest technology, value added food products *etc.* Cultivation of edible mushrooms is one of the most economically viable process for the bioconversion of lingo-cellulosic wastes. Mushroom cultivation is simple, low costing, labor intensive and suitable for rural areas which can provide employment for rural, urban, poor and marginal people in many developing countries. Mushroom cultivation will improve the socio-economic condition of farmers, families and solve employment problems of both literate and illiterate, especially women.

Recently, unemployment is increasing rapidly both in developed and developing countries. In this situation, self-employment can be one important way to increase employment rate for small and marginal poor farm households income and earning extra income. They can easily cultivate mushroom in their home yard because it requires small piece of land. The objectives of rural development in developing countries are mainly diversification of rural income and attaining a competitive structure for agriculture in order to increase job opportunities and development

It has been observed that the employment opportunity has been squeezed in the Government sector due to scarcity of funds as well as imposition of restriction by different financial agencies. In India, 70 per cent of the total population depends upon agriculture and allied sectors directly or indirectly. So self-employment opportunity can be created easily in rural areas through agriculture and allied sectors. In this context, we will discuss some of the projects related to agriculture and allied sectors in detail which provide self-employment to the rural youth. Before going in detail of the individual project, we will discuss some of the general principles for smooth management of any project.

Mushroom farming today is being practiced in more than 100 countries and its production is increasing at an annual rate of 6-7 per cent. In some developed countries of Europe and America, mushroom farming has attained the status of a high-tech industry with very high levels of mechanization and automation. Present world production of mushrooms is around 3.5 million tonnes as per FAO Stat and is over 25 million tonnes (estimated) as per claims of Chinese Association of Edible Fungi. In Indian context, the mushroom production systems are mixed type *i.e.* both seasonal farming as well as high-tech industry. Mushroom production in the country started in the 70s but growth rate, both in terms of productivity as well as production has been phenomenal. In seventies and eighties button mushroom was grown as a seasonal crop in hills, but with the development of the technologies for environmental control and increased understanding of the cropping systems, mushroom production shot up from mere 5000 tonnes in 1990 to over 1,00,000 tonnes in 2010. Today, commercially grown species are button and oyster mushrooms, followed by other tropical mushrooms like milky and paddy straw mushroom *etc.* The nominated areas of production in India are the temperate regions for the button mushroom, tropical and sub-tropical regions for oyster, milky, paddy straw *etc.* Two to three crops of button mushroom are grown seasonally in temperate regions with minor adjustments of temperature in the growing rooms; while one crop of button mushroom is raised in North Western plains of India seasonally. Oyster, paddy straw and milky mushrooms are grown seasonally in the tropical/

sub-tropical areas from April to October. The areas where these mushrooms are popularly grown are Orissa, Maharashtra, Tamil Nadu, Kerala, Andhra Pradesh, Karnataka and North Eastern region of India. Some commercial units are already in operation located in different regions of our country and producing the quality mushrooms for export.

India produces about 600 million tonnes of agricultural waste per annum and a major part of it is left out to decompose naturally or burnt *in situ*. This can effectively be utilized to produce highly nutritive food such as mushrooms and spent mushroom substrate can be converted into organic manure/vermi-compost. Mushrooms are grown seasonally as well as in state-of-art environment controlled cropping rooms all the year round in the commercial units. Mushroom growing is a highly labour-oriented venture and labour availability is no constraint in the country and two factors, that is, availabilities of raw materials

Opportunities

Mushrooms can play an important role in contributing to the livelihoods of rural and urban dwellers, through food security and income generation. Mushrooms can make a valuable dietary addition through protein and various micronutrients coupled with their medicinal properties, mushroom cultivation can represent a valuable small-scale enterprise option. Mushrooms can be successfully grown without access to land and can provide a regular income throughout the year. Growing mushrooms also helps to avoid some of the challenges facing collectors of wild fungi, including species identification, obtaining access and permits for collecting and practicing sustainable harvest. Cultivation is also independent of weather and can recycle agricultural by-products as composted substrate which, in turn can be used as organic mulch in growing other horticultural crops, including vegetables. Mushroom cultivation is highly combinable with a variety of other traditional agricultural and domestic activities and can make a particularly important contribution to the livelihoods of the disabled women and the landless poor who with appropriate training and access to inputs can increase their independence and self-esteem through income generation. However, any interventions to promote livelihood activities should be carefully planned and it is important at the outset to agree with potential mushroom growers: cultivation objectives and the skills, assets and resources available, as well as to identify what market opportunities exist, should they wish to trade their harvested crop.

Attributes of Mushroom Production

☆ Low cost enterprise, requires little infrastructure, space and money.

☆ Most suited for poor/marginal farmers and rural youth in order to raise social status.

☆ Most suited for integrated agriculture farming system.

☆ Maintenance of ecological biodiversity.

☆ Eco-friendly.

☆ Improves nutritional status of citizens. Because its diet enrich with high quality proteins, minerals and vitamins which can be of direct benefit to the human health and fitness.

☆ Ideally suited for diversification programme of agriculture.

☆ Ideal agri-business to generate additional income and employment.

☆ It is a short return agricultural business and can be of immediate benefit to the community.

☆ Potential export oriented enterprise.

Benefits of Mushroom Cultivation

☆ Efficient use of agro wastes.

☆ Unemployed youth/women can start this business with minimal funds.

☆ Suitable for landless/small and marginal farmers.

☆ It helps in rural development and promotes small village industry.

Environmental Impacts

(1) It reduces environmental pollution by bioconversion of vast quantities of organic wastes into mushrooms

Organic solid wastes are biomass, which are resulted annually through the activities of the agricultural, forest and food processing industries. It consist three components mainly *i.e.* cellulose, hemicellulose and lignin. The general term for these organic wastes is lignocellulose and it can be used as substrate for growing mushrooms; otherwise, it would create health hazards.

(2) It recycles spent mushroom substrate (SMS) into organic manure

SMS used for mushroom production and left after harvesting of the mushrooms can be used as compost for soil conditioning. It should be noted that this compost besides being rich in nitrogenous material also contains partly degraded lignocellulosic components.

As a livelihood diversification option, mushroom cultivation has enormous potential to improve food security and income generation, being the fast yielding and nutritious food with great medicinal value. Cultivation does not require any significant capital investment or access to land, as mushrooms can be grown on substrate prepared from any clean agricultural waste material. It can be carried out on a part-time basis, requires little maintenance and is a viable and attractive activity for rural, peri-urban and urban dwellers, in particular women and people with disabilities. Through the provision of income and improved nutrition, successful cultivation and trade in mushrooms can strengthen livelihood assets, which not only reduce vulnerability to shocks, but enhance an individual's or a community's capacity to act upon other economic opportunities. The cultivation of *Pleurotus* mushrooms requires less elaborate technologies. It can be easily adapted in rural areas as it can utilize farm wastes and could be an avenue to solving problems associated with deficiency of proteins, minerals and vitamins. During the last two decades, cultivation of *Pleurotus* mushrooms has become popular worldwide because of their desired attributes. These attributes include: the wide choice of species for cultivation under different climatic conditions, ability to grow on a variety of agricultural and industrial wastes, and their richness in culinary and nutritional values. The other tropical mushrooms *viz.*, paddy straw (*Volvariella volvacea*), milky mushroom (*Calocybe indica*), reishi mushroom (*Ganoderma lucidum*) and black ear (*Auricularia polytricha*) can also be grown at different temperatures in different seasons which needs to be intensified, thus the cultivation of mushroom throughout the year. Awareness and training on mushroom cultivation helped in income generation, nutrient supplement and in profitable marketing. Farmers realized the importance of mushroom and incorporated it in their diet. It also provided an opportunity to strengthen the link between farmers and scientists which helped in technology dissemination and overall development for weaker section.

Economic and Social Impacts

Since mushroom cultivation is very less labour-intensive agro-industrial activity, it could have huge economic and social impact generating income and employment for both women and youth, particularly in rural areas of developing countries like India. Total employment due to mushroom industry in China was over 30 million in 2006. More over only 10 per cent of the employment was from mushroom farming, rest all was from food, beverage manufacturing, trading and management, transport, marketing, wholesaling, retailing, export *etc.* The mushroom industry can also have even broader positive spill-overs, generating complementary employment in areas such as accommodation, restaurant services *etc.* Production and consumption of mushrooms have tremendously increased in India mainly due to increase in awareness of commercial and nutritional significance of this commodity. At present production of mushrooms has crossed lakh tonne with annual growth rate of above 15 per cent. The local mushroom industry can also be the main source of revenue generation for government. The following statements summarise the significance of mushrooms towards alleviating poverty, enhancing human health, and arresting environmental degradation:

1. Mushrooms can convert lignocellulosic waste materials into a wide diversity of products, which have multi-beneficial effects to human beings, *e.g.*, as food, health tonic and medicine, as feed, fertilisers, and for protecting and regenerating the environment. In addition, mushroom cultivation can positively generate equitable economic growth. The tropical regions, particularly, have a wet and warm climate and have an abundant collection of agricultural wastes. These materials are resistant to natural biological degradation because they contain mainly cellulose, hemicellulose and lignin. Mycelia of mushrooms can excrete enzyme complexes which can directly attack/degrade these components of lignocellulosic materials. Therefore, mushrooms can use these wastes as nutrients for its growth besides being food and medicine for human consumption.

2. Mushrooms are relatively fast growing fungi. Some tropical mushrooms can be harvested and consumed within 10 days after spawning. By the use of different varieties in combination, mushrooms can be cultivated around the year.

3. Mushroom cultivation can be less labour intensive. Its cultivation is easy in nature and can be easily internalized under house hold daily routine.

4. Agricultural land will decrease due to industrialization, population pressure and converted into wasteland due to intensive agriculture, the mushroom cultivation requires relatively little space, even it is cultivated in air, since the entire cultivation is an indoor affair hence it is highly women friendly in nature.

5. Mushrooms have been accepted as human food since times immemorial, and can immediately supply additional protein to human food.

6. Edible mushrooms should be treated as healthy vegetables. After improving the cultivation techniques, it should be cultivated widely and cheaply like other common vegetables, for the access to every man.

7. This is an enterprise, which does not require much capital investment and skilled labour. Therefore, it can be well adopted by poor, small and marginal farmers, farmwomen, landless labourers, rural unemployed youth and even retired or in-service personnel.

8. In view of its pleasant flavour, high protein content, tonic and medicinal values, mushrooms represent one of the world's greatest untapped resources of nutrition and palatable food for coming age humans.

Chapter 22

Marketing, Export and Financial Services in Mushroom Business

Marketing is getting the right product, to the right people, at the right price, at the right time and in the right way. Marketing of fresh mushrooms in India is not very organized However, trade in the processed product (canned and dried) of mushroom is substantial and organized. For effective and efficient marketing, especially export, it is necessary to understand the global trade vis-à-vis the sources of supply, potential regions of demand and consumption patterns.

Marketing of mushrooms in India is yet to be fully organized. It is simply a system of producing and selling directly to retailer or even to consumer, and suffers from various limitations. Like other countries, where 10 per cent of the total cost is earmarked for marketing, we have not given the similar thought to the mushroom marketing. Per capita consumption of mushrooms in India is less than 50 g as compared to 1 kg in various other countries. Not much serious effort to promote the product, to strengthen and expand the market has been made. Mushroom is a novel food item and it is known for its flavour, textures, nutritive value. But many of Indians are not aware of 'what is mushroom and if aware, they are still confused whether it is a vegetable or non-vegetable?' The popularization of deliciousness, nutritive and medicinal value of mushroom by advertising, literatures, posters and demonstrations may be highly helpful to aware the people about mushroom consumers. The demand of processed and fast foods is surging day by day and mushroom hold much scope and potential. Mushrooms may be canned to meet the demand in the off-season and in nonproducing areas. Product diversification is being tried. Research is being done to bring down the cost of production of mushrooms and processing to promote the sale/export of canned mushrooms, in order to compete in the international market. There is not much problem in the sale of fresh *Pleurotus* due to very low production cost but there have been problems in selling dried 'Dhingri'. The chain of marketing is not much straight and therefore, there is no direct linking between the exported quantity and production by growers. Formation of cooperative society and federations is advisable to make a continuous, direct and established market of mushroom. *Plerutous* growers may form a cooperative where they may pool their product and trade. Agricultural and Processed Food Products Export Development Authority (APEDA), Central and State agencies are eager to help them once they are assured of sufficient consignment for export, for 2-3 years by the society or federation. The study of obstacles in marketing of mushrooms may help to tackle and expand the market.

Stategies for Successful Markeing of Mushroom

1. Effective storage facility: It includes a. Creation of cold storage facility. b. Creation of refrigerated transport facility. c. Creation of processing facility.
2. Lowering the cost of production and bringing down the sale price to boost the demand.
3. Good pre-packing for eye appeal.
4. Guidance to retailers regarding handling, storage, food value and recipes.
5. Approach of supermarkets, chain vegetable stores, general store, mother dairy retail counters for retail sale.
6. States marketing policy.
7. Initiative of Public sector, processing and export organizations in marketing.
8. Assured supply throughout the year at a reasonable constant price is a key to good marketing. Efforts should be made to diversify and cultivate different mushrooms throughout the year along with cultivating some of the important mushroom during off-season under controlled condition.
9. In a limited area, say a village or a cooperative, the crops should be time-scheduled to get a daily reasonably uniform production to avoid glut on a day, this is required to meet the commensurate demand. One cannot ask consumer to purchase more because all have got a peak flush during a week. The marketing system has to be viewed as a value chain where all components are taken care of. The increased production should get translated into economic gain for mushroom producers. This is possible only when marketing is organized and demand is ensured. Considering the awareness about consumption and health benefits of a number of mushrooms available for cultivation is limited, their demand is also less. Hence, multipronged strategy is required that means the need to expand market, increase demand, organize marketing and form cooperatives.

Market Research

The main aim of market assessment is to assess the choice and demand of consumers, product competition, production cost, price flow and determination of retail cost and bulk cost of the product. Many posh quality of mushrooms can be cultivated, but the market is yet to be expanded. Therefore, it is advisable and wise worthy to assess the market and consumption of produced mushroom before investing large amounts of capital in the enterprise. Few fundamental and preliminary methods to conduct marketing research are;

☆ Observation of buyers.

☆ Surveys of stores.

☆ Individual interviews with growers.

☆ Marketing analysis (once you have an experimental product).

☆ Competition Assess.

This will help you to know the existence and status of mushroom market and will give you an idea to invest and engage the cost of production. It is good enough to know the quality and consumption of the product of other competitors for continuous economic improvement of your product and to stay and excel in the market competition.

Market Channels

Explore as many marketing strategies as;

☆ Market the fresh or dried product directly to your customers (Door to door).

☆ Farmer to big stores, hotels.

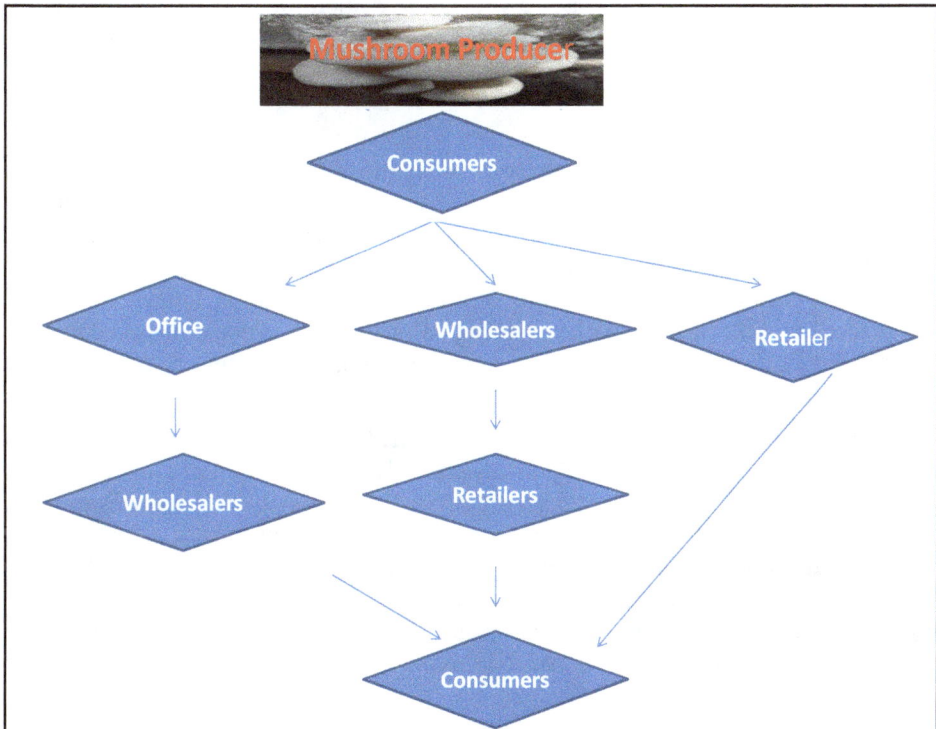

Figure 22.1: Marketing Channels of Mushroom Business.

☆ Farmer to local market.

☆ Farmer to distributors.

Direct Marketing

If you can sell your mushrooms or its products directly to an end user, you will naturally receive a better price than if you sell to a wholesaler. Direct marketing of mushrooms at local farmers' markets, to restaurants, to hotels or in supermarkets is possible in many locations. When competing in local markets, excellent service, top quality, and consistent supply is necessary. Some chefs specialize in locally grown foods and may be interested for that reason. Others are willing to pay for fresh, premium produce. Local grocery stores are another potential buyer of fresh mushrooms. Natural foods stores are a market that may be more tolerant of seasonal supply. Any grocer will require assurances of both quality and regular supply before switching from established wholesale sources. Although the wholesaler with an established account creates taut competition, the small, efficient producer might still have an advantage in some niche markets. For instance, shiitakes grown on logs are generally of higher quality and have a longer shelf life than shiitakes grown on sawdust substrates (the most common mass-production method). Log-grown shiitakes earn prices from three to eight times higher than those grown on sawdust substrates. Find the buyer to whom, quality matters, and you will have found a market for your product. Locally grown oyster mushrooms have an advantage because oysters have a very limited shelf life and are too delicate to transport easily. The grower with direct, local sales can supply a fresher product that arrives in better condition.

Financial Services

Fortunately small-scale mushroom cultivation does not require significant financial assets to establish an enterprise. Cash, savings and access to credit or grants are seldom essential to initiate small scale cultivation systems, sufficient to provide a nutritious source of food and reliable source of instant cash. Financial resources will however become more important as the size of an enterprise scales-up, or if cultivators want to explore adding value through processing and consider investment in drying equipment, or secure specialist containers to package and transport products further to more distant markets.

The types of credit available vary from Central and state governments and between State to State. Private organizations and banks are normally good sources of credit for establishing mushroom business. Farmers will raise cash from farm gate sales or from agents or cooperatives marketing their produce. Cooperatives are often in a better position to offer credit to rural farmers than individuals or financial institutions. External funding can be used to provide more efficient or higher technological processing equipment, facilitate information and exchange visits, and provide training to expand cultivation skills.

Chapter 23

Mushroom Cultivation Management

Mushroom cultivation technology is the discipline that deal with the principles and practices for economic mushroom cultivation. Mushroom cultivation practice requires understanding of many avenues of management for implementation of techniques and resulting economic production. Management plays a decisive role right from the establishment of laboratory to the marketing. Therefore, it is essential to understand the management involved at every step of the mushroom cultivation and marketing. Different important factors of mushroom cultivation management have been dealt here.

☆ Selection of site for mushroom house

☆ Care in laboratory condition

☆ Selection of the mushroom strain and maintenance

☆ Spawn

☆ Type of substrate used

☆ Spawning

☆ Mushroom house

☆ Management of pest, diseases and disorders

☆ Harvesting and picking of mushroom

☆ Postharvest technology and marketing.

Selection of Site for Mushroom House

Site of construction of any laboratory/house depends on the availability of other components which helps in the purpose it has been established for. Selection of site/location for the establishment of the mushroom farm requires considerations such as availability of the substrate, preferred species, suitable species as per the environment, environmental conditions, electricity, transport facilities, market opportunity, available technology and the acceptability of the specific mushroom to the local people *etc.*

Care in Laboratory Condition

Proper basic knowledge, precautions, working principles of the equipments and information regarding nature of the chemicals are prerequisites for commencing any laboratory work. Spawn laboratory and mushroom house/cropping room are two essential subunits of mushroom production unit. Proper maintenance of the mushroom production unit and knowing of essential laboratory practices determine the success and production of mushroom. Every operations from pure culture to spawn preparation must be conducted under ambient conditions and contamination free environment. It is advisable to follow good laboratory and safety practices while working in laboratory.

Production of the Mushroom Culture or Starter

The first stage in any mushroom cultivation process is to obtain a pure mycelial culture of the mushroom strain of interest. Large scale growers create own facilities to produce their strains themselves, either thorough single or multispore cultures or by tissue culture of a high yielding and vigorous strain. Many strains have already been developed through genetic breeding programs. Culture of each type of mushroom generally requires specific substrate formulation for growth and maintenance of purity.

Spawn

Spawn is the planting/seed material and quality of spawn determines the cultivation efficiency of mushroom. Therefore, growers need to use good quality spawn for commercial production. Most of the mushroom growers do not produce spawn themselves, as it is a very complicated process. Few companies are engaged in the production of spawn by inoculating grain with spores. Procurement of spawn from company is supposed to be convenient for growers. Old spawn is not acceptable because old spawn lose its vigor with the passage of time. Therefore, it is advisable to know the storage potential and its probable duration of viability. Vigorous growth of the commercial spawn is a prerequisite to good growth and yield. If, the spawn is not vigorous, competent organisms will overgrow to mushroom mycelia. The quantity of the spawn used does not directly affect yield. The greater the amount of spawn used, the faster it will colonize in the substrate. As a result, the growths of competitors are hindered, and yield remains unaffected. 2-4 per cent of spawn is suggested to be sufficient for inoculation into the substrate in case of most of the species of mushroom. Once the bottle is opened, entire spawn must be used. However, unused and opened bottles or bags of not contaminated spawn can be kept in the refrigerator for 2 to 6 days during storage.

Type of Substrate Used

The most commonly mushrooms consumed are saprophytes, decomposers and grow on organic matters like wood, leaves and straw in nature. Mushroom requires carbon, nitrogen and inorganic compounds as its nutritional sources. Cellulose, hemi-cellulose and lignin are used as a key substrate as a carbon source, therefore, most organic matters containing cellulose, hemi-cellulose or lignin can be used as a substrate such as; cotton, cottonseed hull, corncob, sugarcane waste, sawdust, and others. However, required quantity of nutritional source differs from species to species, as biomass composition differs. For instance, button mushroom (*Agaricus bisporus*) requires relatively high nitrogen source while oyster mushroom and shiitake require less nitrogen and more carbon. Mushroom mycelia secrete digestive enzymes on the substrate and absorb digested macromolecules. Mushroom is also influenced by the biochemical environment of substrate. The optimal pH value of substrate should be in between 6 to 8.

Spawning

Spawning is the inoculation of the culture into the substrate or compost and the method of spawning greatly affect the production. It requires diligent handling and strict follow of procedure. Spawning is followed by spawn run throughout the substrate. Mushroom mycelium starts to produce more enzymes and break down the macromolecules of the substrate rapidly and absorb simple molecules into the mycelium for further growth and development into mushroom. This period of spawn run is also called the vegetative growth or inoculation period and the length of spawn run vary from species to species. *Agaricus* and *Pleurotus* require 2-4 weeks for spawning, *Volvariella*, 10-15 days and *Auricularia* and *Lentinula edodus* 2-3 months.

Mushroom House Management

Design of Mushroom House

Mushrooms are grown in shade whether it is concrete buildings or thatched house (mushroom hut). Mushroom cultivation require two rooms; one spawn running room for spawn run/mycelial growth and other cropping room for cropping/fruiting. Both rooms require different environmental condition for growth which vary from species to species. There is no standard size or design of buildings or thatched house for mushroom cultivation, only construction costs, machinery space requirements, tray or bed size, and stacking design should be taken under consideration. Doors must be designed to ensemble all machinery and equipment that is used. Mushrooms do not require complete darkness to grow but also direct sunlight to reach the beds. Buildings should be rodent and snake proof. Cement floors with adequate drainage are required to allow for easy cleaning and hygiene. Roof of concrete buildings should have sufficient slope to prevent condensation and dripping onto beds. Insulation (commonly polystyrene panels) prevents temperature fluctuations and increases the efficiency of the air conditioning. Good ventilation to supply a constant flow of fresh air and prevent carbon dioxide buildup is essential. Ventilation units should be fully adjustable in terms of circulation volumes, and should include a filter to prevent entry of insects and airborne spores of other fungi.

Figure 23.1: Arrangement of Mushroom Bags in the Racks/Shelves of Bamboo in Mushroom House.

Figure 23.2: Arrangement of Mushroom Bags in the Racks/Shelves of Cement in Mushroom House.

Figure 23.3: Arrangement of Mushroom Bags in the Air Space in Mushroom House.

Figure 23.4: Artificial Rain Water on Mushroom Bags.

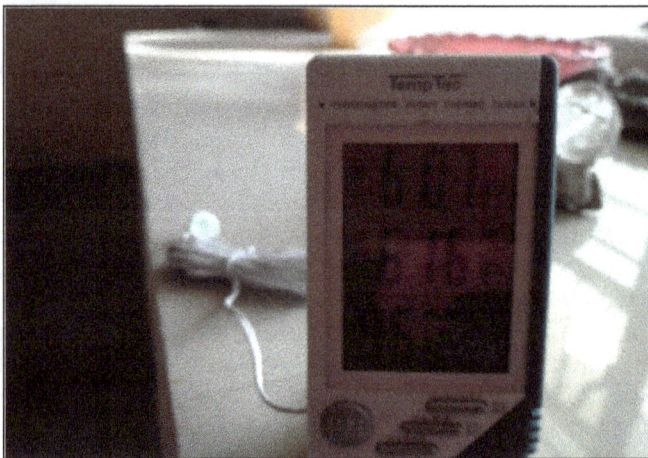

Figure 23.5: Hygrometer Device which Observe the RH in the Mushroom House.

The filters should be cleaned regularly. Recycling of unfiltered air between different growing rooms is not good. Shelves/racks should be placed in the mushroom house with proper distance and height. In addition, the shelves/racks should be placed at right distance and convenient to manage all activities such as watering, handling, aeration, harvesting *etc*. Production containers used for mushroom cultivation are trays and bags. Trays are supposed to be most suitable for the cultivation of button, milky and reishi mushroom, while bags (poly propylene bag) are most suitable for the cultivation of oyster and paddy straw mushroom. Trays and bags should be set on shelves/rack with sufficient space to allow easy air circulation. Wood, concrete, or rust-protected metal may be used for the construction of trays, because these are sturdy enough to withstand rough handling. Plastic bags are not reusable and cheap, but require extra care during watering. Maximum space can be utilized holding bags in the air space with the help of rope. The thatched (mushroom hut) house is more economic, easy to make and require less investment over concrete building. This advantage of thatched house is encouraging for the low cost investee or grower. However cultivation in thatched house attracts more pest and snake and intense care must be taken to keep it off.

Growth Environment

Controlled environment is required in the mushroom house for efficient production of high-quality mushroom. Temperature, humidity, light and aeration are important environmental factors which affects the fruiting and growth. Designing and selection of materials required in the construction should be as per the required temperature and humidity or nearer. Required environmental condition may differ from species to species and stage to stage (mycelia run, pin head formation and fruiting) in a single species. Air conditioner, blower and humidifiers are set in the house to control optimum temperature and Humidity. Measuring devices such as thermometer and hygrometer are set in the centre of the mushroom house to record the optimum temperature and humidity during the production cycle. Temperature and humidity may also be controlled by the light spray of water as per the existing condition. Therefore, management of temperature, humidity, aeration are very decisive in the cultivation room as prescribed for each stage, right from spawning to harvesting.

Management of Pest and Diseases

Diseases, pests and disorders affects mushroom production to a great extent, therefore, management of those diseases and pests is very important for economic mushroom production. Occurrence of diseases and infestation of pests in mushroom cultivation takes place due to following reasons:

☆ High humidity and warm temperature favours many pathogens and pests.

☆ Chemical control for diseases or pests has its own limitation.

☆ Pathogens and pests are readily attracted inside and/or outside mushroom houses involved with continuous cultivation.

☆ If growing houses are not well equipped with environmental control than more chances of pest and disease attack.

Maintenance of hygienic condition play very important role in keeping pest and disease off (for detail refer chapter 20). Every practice must be oriented to exclusion and elimination of pathogens or pests as far as possible.

☆ Doors should be closed immediately after finishing all operation.

☆ Any practices that expose substrates to pathogens or pests during bagging and spawning must be avoided.

☆ Entry of mushroom flies into mushroom houses must be prohibited by installing screens on windows and doors.

☆ Mushroom house/cropping room should be inspected regularly and contaminated materials must be discarded immediately

☆ Mushroom bags or beds must be cleaned by removing any mushroom debris or mushroom stumps shortly after harvest.

☆ Floors should be clean. Dumping of waste near mushroom houses, attract mushroom flies.

☆ Spent substrate must be disinfected or pasteurized before removing it from mushroom houses after cultivation.

☆ Infected mushroom in the houses must be discarded thoroughly before entering new bags and trays.

☆ Clean and disinfected equipment only must be utilized in the mushroom house.

☆ Worker should wear a laboratory coat, a mask, a lab slipper and a pair of hand gloves while working.

Harvesting and Picking of Mushroom

After full mycelium development, when the substrate has been fully colonized or penetrated by the mushroom mycelium, mushroom begin to form, first as pinhead primodia, then develop into buttons and finally into umbrella-stage mushroom called mushroom fruit. Time of harvesting is an important consideration as mushroom grows quickly, doubling their size within 24 hours. The fruiting bodies are harvested by hand by grabbing from the base and performing a twisting motion. Although hand picking require more time, but it offers the best guarantee that the mushrooms will be removed from the beds undamaged. On an average, a picker can harvest between 18 and 30 kilos of mushrooms an hour. The stems are trimmed. Pulling the mushroom straight out of the mycelium with too much force can damage the mycelium and the harvested mushroom is usually graded straight into boxes for transport and sale. Nine days after the first flush, the second flush will be harvested. After harvesting, you can use a small brush to gently brush off any vermiculite or substrate left on the mushroom. However, do not wash the mushrooms. The second flush often consists of larger, but less mushrooms yield than the first flush. *Agaricus, Calocybe* and *Volvariella* are usually picked at the button stage while *Pleurotus* and *Auricularia* are picked once the fruit has opened. Harvested mushrooms need to be carefully handled and should be kept in a container that allows air circulation, such as a basket. It also needs proper care to prevent it from staining. The baskets

Table 23.1: Problems, Causes and their Solution during Mushroom Cultivation

Sl.No.	Problem	Cause	Solution
1.	Mushrooms taking too long to appear after bag is opened.	Mycelium not sufficiently mature. Fluctuation of temperature and humidity, Spawn too weak or degenerated.	Allow proper maturing of spawn. Maintain proper temperature and humidity.
2.	Mushroom fruits are small.	Spawn fragile or degenerated. Nutrients deficiency. Too many fruits developing at the same time.	Use reliable spawn. Provide sufficient nutrients. Allow only a few fruits to develop at one time by opening bags slowly.
3.	Mushrooms decaying before picking.	Fungal or bacterial growth. Excessive watering.	Discard mushroom or bag to prevent spread of fungal or bacterial infection. Avoid direct watering on developing fruits.
4.	Too few fruits harvestable.	Insufficient supplement on the substrates, week or degenerated spawn.	Provide sufficient nutrients. Use reliable spawn.
5.	Failure to form fruit body.	Spawn old or degenerated. Poor ventilation and high temperature	Use reliable spawn. Maintain proper temperature and ventilation.
6.	Mycelium running but no fruits formed.	Degenerated spawn.	Use reliable spawn.
7.	Mycelium or spawn not growing on the substrate.	Old/weak/dead spawn. Temperature too high or low. Contaminated or old substrate.	Use good quality spawn. Use freshly harvested straw.

containing mushrooms should be covered to keep flies out and to protect from sunlight, high temperatures and droughts. Harvested mushrooms either should be taken to market without delay in order to maintain their freshness and quality, or stored in a refrigerated environment or processed for value addition. Yield of mushroom is influenced by many parameters; substrate used, compost depth and quality, length of cropping and grade of mushrooms picked, spawn productivity, moisture and climatic conditions, and disease factors. Therefore, it is important to take the every parameter into consideration right from culture to the marketing.

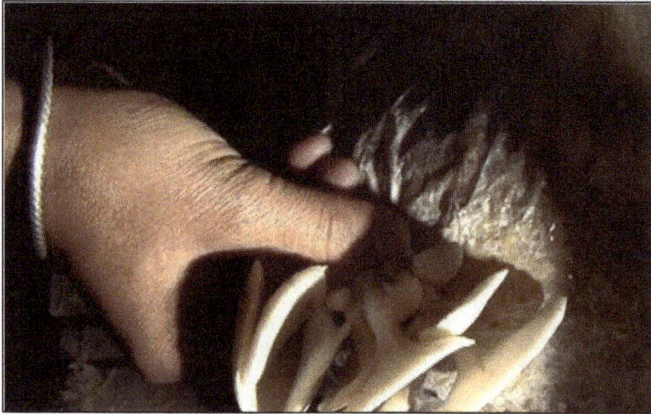

Figure 23.6: Harvesting and Picking of Mushroom by Hand with Twisting Motion.

Post Harvest Management

Fresh mushrooms can be stored in the refrigerator, but only for 3-10 days (depending on the species and moisture content). Storage for long duration, mushroom drying is better option. Mushrooms contain about 90 per cent water, so drying them is not only an excellent method for storage, it also decreases their size, thus making transport easier. To dry mushrooms, all you need is air circulation and a warm spot. An effective method for drying mushrooms is by placing them on kitchen paper and directing a fan at the mushrooms to provide a constant airflow. At a temperature of 25-30°C the mushrooms should be cracker dry in a few days. There are several ways to store dried mushrooms are mentioned detail in **Chapter 18**.

Glossary

Agar media: It is a solidifying media which is prepared from seaweed. It usually comes in powder form commercially.

Ascocarp: The fruiting body of ascomycetes bearing or containing asci.

Ascomycetes: A group of fungi producing their sexual spores, *i.e.*, ascospore within asci.

Ascospore: A sexually produced spore borne in ascus.

Autoclave: This equipments is used to sterilize heat and steam resistant materials under high pressure saturated steam. Generally the pressure of 15 psi is applied at 121°C (249 °F) for 15–20 minutes to sterilize the materials. However these parameters may vary depending on the size of the load and the contents.

Basidiocarp: Fruiting body of basidiomycetous fungus.

Basidiomycetes: A group of fungi producing their sexual spores, *i.e.* basidiospores on basidia.

Basidiospore: A sexually produced spore borne on basidium.

Biological Oxygen Demand (BOD): It maintains a range of temperature below and above the ambient temperature required for growth and multiplication of microorganism.

Buttons: Mushroom with closed cap with membrane, when stem length does not exceed to 2 cm.

Carpophore: This is fleshy, spore producing body of basidiomycetes and ascomycetes. Commonly, the term mushroom is applied to carpophores which have a distinctive stipe and cap.

Casing: Casing means covering the top surface of compost bags after spawn run is over to stimulate pinning and fruiting of mushrooms. Thickness of casing material is about 2-3 cm. Casing provides physical support, moisture and allow gasses to escape from the substrate.

Compost: A mixture of decomposed organic and inorganic substances with nutrient composition selective for the growth and fructification of the common cultivated mushroom.

Composting: It refers to the piling up of substrates for a certain period of time and the changes due to the activities of various micro-organisms, which result in the composted substrate being chemically and physically different from the starting material. This is sometimes referred to as a solid state fermentation.

Culture: Microorganism, usually *in vitro* on a prepared medium and artificially maintained on such food material.

Cup-fungi: Fungi with cup shaped fruiting body.

Cups: Mushrooms with well developed membrane and just opening with cap retaining a pronounced cup shaped, when stem length does not exceed 2.5 -7.0 cm.

Dikaryotic mycelium: This is mycelium containing two sexually compatible nuclei per cell and is also called secondary mycelium. Only dikaryotic mycelium can give rise mushroom.

Flush or breaks: The collective formation and development of mushrooms within a short time period, often occurring in a rhythmic manner. The speed of growth and size of the flushes varies per type of mushroom.

Flushes: The appearance of mushrooms normally occurs in rhythmic cycles called "flushes".

Fructification: Production of spores by fungi; also a fruiting body.

Fructification: The act of fruit body formation.

Fruit body: The sexual reproductive body of the mushroom.

Fruiting body: It is spore bearing structure or bear fungus spores.

Fruiting culture: It is defined as a genetic capacity of culture to form fruiting bodies under suitable growth conditions.

Gill trama: Gill tissue lying between two hymenium layers.

Gill: The structure resembling plates bearing hymenium of basidiomycetes.

Grain spawn: Grain is uses a substrates for making spawn.

Granular spawn: Shell powder, starch, compost powder and grain hull powder is used for making granular spawn.

Heterotrophs: Organism that cannot synthesize its own food and remain dependent on complex organic substances for nutrition are called heterotrophs.

Hybridization: Hybridization is the process of interbreeding between individuals of different species.

Hymeniaum (pl. Hymenia): The layer of fertile spore bearing cells on the gill.

Hypha: (pl. Hyphae) Individual cells of mycelium which is filamentous or thread like.

Hypogeous: Organism growing underground is known as hypogeous.

Lamella: A plate like structure (gill) on which some basidiomycetes produces their basidia.

Lamellate: Having gills.

Laminar air flow: Laminar flow machine provide aseptic or microorganism free environment for performing various activities such as pouring of sterilized media, isolation, inoculation and pure culture of mushroom.

Manual harvesting: Mushrooms are harvested by hand.

Mechanical harvesting: Mushrooms are harvested by mechanically.

Microscope: A microscope is an instrument used to see objects that are too small for the naked eye.

Monokaryotic mycelium: A spore from a mushroom which germinate starts with the formation of primary mycelium. This mycelium is also called monokaryotic mycelium. monokaryotic mycelium on its own cannot forms mushroom.

Mother spawn: Mother spawn prepared using pure culture mycelium is placed onto steam-sterilized grain, and in time the mycelium completely grows through the grain.

Multispore: A multi strain culture that contains an array of spores that could each develop into different unique mushroom with their own traits

Mushroom substrate: It may be simply defined as a lingo-cellulosic material that supports the growth, development and fruiting of mushroom mycelium.

Mushroom: Mushroom is fleshy, fruiting body of macroscopic fungi

Mycelium (p1. mycelia): A network of hyphae.

Mycology: The study of fungi.

Mycophagist: A person or animal that eats fungi.

Mycophagy: Eating of mushroom.

Mycophile: A person who likes mushrooms.

Oyster mushroom: Oyster mushroom is scientifically known as *Pleurotus* or Dhingri mushroom in Hindi.

Pasteurization: This is short time exposure of heating, to kill viable pathogens so that there may be less chances of contamination.

Permeated compost: Compost that has been mixed with grain spawn. The mycelium permeates the compost. The grower creates the perfect conditions under which the mycelium will start sprouting mushrooms.

Petri dishes: It is used for cultivation of mycelium on agar.

Pinheads: Emergence of structure from casing layer is called pinheads, which give rise mature mushroom.

Planting spawn: Also called commercial spawn in which poly propylene (PP bag) is used as a containers for filling grains.

Plasmogamy: This is the fusion of cells or protoplasts but not of nuclei. This event commonly takes place in higher terrestrial fungi including mushroom.

Primodia formation: It is the fruiting stage of mushroom after the emerging of pinheads.

Rhizomorph: A cord like strand of fungus hypae.

Saprophytes: Organism which grow and derive nutrient from dead or decaying organic matter are called saprophytes.

Saw dust: It is the by-product of cutting, grinding, drilling, sanding, or otherwise pulverizing wood or any other material with a saw, which is used for cultivation of different kinds of edible mushroom.

Sectoring: Sectoring is any type of mycelial growth that differs in appearance, growth rate and colour from the typical appearance of a given strain.

Slant: These are test tube filled with agar medium for mycelium to grow.

Spent mushroom substrate (SMS)/spent compost: Spent mushroom substrates (SMS) is the substrate left after harvesting mushroom fruit bodies.

Spawn (Mushroom seed): A steam-sterilized grain colonized by the mycelium of the mushroom and used to "seed of mushroom. Spawn is the vegetative mycelium from a selected mushroom grown on a convenient medium like wheat, pearl millet, sorghum, etc for raising mushroom crop or mushroom spawn is the mushroom mycelium growing on a given substrate.

Spawn run: Spawn run is the period when the mycelium grows from the spawn to cover all of the substrate.

Spawning: The process of mixing spawn with substrate is called spawning.

Specialty mushrooms: It is a term given to a group of mushrooms, that are not grown commonly and available commonly in the market.

Spore print: A spore print is a piece (tinfoil or glass slide) used to collect the spores of a particular mushroom strain. Professional mycologists use a spore print for identification of mushroom species and store its released spores.

Spore syringe: This is like a syringe with spores of a mushroom in sterilized water.

Spore: A single or many celled reproductive unit, as of fungi, which may germinate and develop into a new individual.

Stipe: The stem or stalk of the mushroom.

Substrate: The organic materials (A nutrient rich materials), from which mushrooms derive their nutrition, are referred to as substrates.

Supplementation at casing: The addition of nutrients to colonized compost at time of casing application.

Supplementation at spawning: The addition of delayed release nutrients at time of spawning.

Thallus (p1. Thalli): Structure which is undifferentiated root, stem and leaf.

Toadstool: A kind of fungus that is similar to mushroom, that consist of a round cap on a short stem that is often poisonous or inedible.

Auxiliary Information

A. Books

a. English

1. Bahl, N. (1984). Handbook on mushroom, Oxford and IBH, New Delhi
2. Garcha, H.S. (1984) A Manual of mushroom growing, PAU publication, Ludhiana, Punjab.
3. Kannaiyan, S. and Ramasamy, K. (1980) A handbook of edible mushroom. Today and Tomorrow Printers and Publishers, New Delhi.
4. Marimuthu, T., Krishnamoorthy, A.S. and Jeyarajan, R. (1991) Glimpses of mushroom research in Tamil Nadu, TNAU Publisher, Coimbatore.
5. Pathak, V.N., Yadav, N. and Gaur, M. (2013) Mushroom production and processing technology. Agrobios, Jodhpur.
6. Purkayastha, R.P. and Chanda, A. (1985) Manual of Indian edible mushroom. Today and Tomorrow Printers and Publishers, New Delhi.
7. Sharma, S.R. and Mehta, KB. (1991) Bibliography of mushroom research of India, NCMRT Publication, Chambaghat, Solan.
8. Tiwari, S.C. and Kapoor, P. (1988) Mushroom cultivation: An economic analysis. Oxford and IBH, New Delhi.

b. Hindi

1. Gupta, Y. and Vijay, B. (1992). Shwet button Khumb Ka Utpadan. NCMRT Publication, Solan, H.P.
2. Kumar, S. and Chand, G. (2014). Mushroom ki Vyvsaic Khrti. Directorate of Extension Education, Bihar Agricultural University, Sabour, Bhagalpur, Bihar.

3. Kumar, S. and Chand, G. (2014). Mushroom Utpadan. Directorate of Extension Education, Bihar Agricultural University, Sabour, Bhagalpur, Bihar.

4. Kumar, T.R., Shandilya and Chaudhary, R.P. (1989). Kuumb ki kheti (Mushroom cultivation). Directorate of Extension Education, U.H.F., Solan.

5. Pathak, V.N., Singh, R.B., Majumdar, V.L. and Verma, O.P (1983). Mushroom ki Kheti. SKN College of Agriculture, Jobner.

6. Ram Chandra (2014). Mushroom Utpadan: Takneek awam Vavsaikarad, Department of Mycology and Plant Pathology, BHU, UP.

B. Major Jornals of Mushroom

1. Indian Journal of Mushroom, Indian Mushroom Growers Association, Mushroom Research Laboratory, Chambaghat, Solan-173213, H.P.

2. Mushroom Research, Mushroom Society of India, Directorate of Mushroom Research (DMR), Chambaghat, Solan-173213, H.P.

3. Journal of Mycology and Plant Pathology, Indian Society of Mycology and Plant Pathology, Rajasthan College of Agriculture, Post Box 154, Udaipur 313 001, Rajasthan, India.

4. Journal of Mycopathology Reseach, Indian Mycological Society, Kolkata, W.B., India.

C. Collection and Conservation of Mushrom Cultures

1. Directorate of Mushroom Research (DMR), Chambaghat, Solan-173213, H.P.

2. Indian Type Culture Collection (ITCC), IARI, New Delhi-110012.

3. Herbarium Cryptogamie India Orientalis (HCIO), IARI, New Delhi-110012.

D. References

1. Directorate of Mushroom Research, ICAR, Chambaghat- 173213, Solan, Himachal Pradesh, India.

2. Thakur, M.P. (2014). Present status and future prospects of tropical mushroom cultivation in India: A review. *Indian Phytopath.* 67 (2): 113-125.

3. Prakasam, V. (2012). Current scenario of mushroom research in india. Indian *Phytopathology,* 65 (1): 1-11.

4. Singh, R.P. and Mishra, K.K. (2013). Mushroom cultivation. Mushroom Research and Training Centre, G.B. Pant University of Agriculture and Technology, Pantnagar, Uttarakhand, pp. 2-25.

Appendix

Table 1: Requirement of Appliances/Equipments, Glasswares, Chemicals and Miscellaneous Items in the Mushroom Laboratory

Sl.No.	Appliances/Tools	Sl.No.	Appliances/Tools
	A. Laboratory appliances/tools		
1.	Autoclave	2.	Laminar air flow/Inoculation room
3.	BOD Incubator/Incubation room	4.	Refrigerator
5.	Hot plate	6.	Pan (different size)
7.	Sprit lamp	8.	Inoculation needle
9.	Forceps	10.	Knife/Blade
11.	Scissors	12.	Cork borer
13.	Scalpel	14.	Glass maker
15.	Gel electrophoresis	16.	Centrifuge
17.	Microscope	18	Spatula
19.	Thermometer	20.	Humidifier/Hygrometer
21.	Mesh (For filter water from substrate-3x6 fit	22.	LPG/Electric stove (for boiling substrates)
	b. Glasswares		
1.	Conical flasks (different size)	2.	Beaker (different size)
3.	Petri dishes	4.	Culture tube (different size)
5.	Measuring cylinder (different capacities)	6.	Pipette (different volume)
7.	Cover-slip	8.	Glass slides
9.	Bearman funnel		

Contd...

Table 1–*Contd...*

Sl.No.	Appliances/Tools	Sl.No.	Appliances/Tools
	C. Chemicals		
1.	Calcium carbonate	2.	Calcium sulphate
3.	Agar-agar	4.	Dextrose
5.	Malt extract	6.	Formalin
7.	Spirit	8.	Alcohol 70 per cent
9.	Bavistin	10	Washing powder
	D. Miscellaneous items		
1.	Non-absorbent cotton	2.	Aluminium foil
3.	Rubber band	4.	Blotting paper
5.	Washing brush	6.	Wash bottle
7.	Wire basket	8.	Mortar and pestle

Table 2: Types of Mushroom and its Common Name

Sl.No.	Spp. of Mushroom	Common Name
1.	*Agaricus bisporus*	Temperate button mushroom
3.	*Agaricus bitorquis*	Summer white button mushroom
4.	*Pleurotus ostreatus*	Gray oyster mushroom
5.	*Pleurotus florida*	White mushroom
6.	*Pleurotus sajor-caju*	Grey mushroom
7.	*Pleurotus d'jamour*	Pink mushroom
8.	*Pleurotus cornicopiae*	Golden mushroom
9.	*Pleurotus eryngi*	King oyster
10.	*Pleurotus cystidiosis*	Ohritake mushroom
11.	*Calocybe indica*	White milky mushroom
12.	*Volvariella volvacea*	Paddy straw/Chinese/Tropical mushroom
13.	*Volvariella diplasia*	Banana Straw mushroom
14.	*Lentinula edodus*	Shiitake mushroom
15.	*Stropharia ruguso-annulata*	Giant mushroom
16.	*Flammulina velutipes*	Winter mushroom
17.	*Tremella fuciformis*	Silver ear mushroom
18.	*Morchella* sp.	Morel (Gucchi) mushroom

Medicinal mushroom

1.	*Lentinula edodus*	Shitake mushroom
2.	*Ganoderma lucidum*	Reishi mushroom
3.	*Cordyceps sinensis*	Chinese caterpillar mushroom
4.	*Auricularia polytricha*	Black ear/wood ear mushroom
5.	*Grifola frondosa*	Maitake (means dancing mushroom)

Poisonous mushroom

1.	*Amanita phaloides*	Death cap
2.	*Amanita muscaria*	Fly agric
3.	*Amanita pantherina*	Panther cap
4.	*Amanita citrina*	False death cap
5.	*Lactarius* sp.	Milk caps
6.	*Coprinus atramentarius*	Inky cap mushroom

Pathogenic mushroom

1.	*Auricularia auricula*	Black ear mushroom
2.	*Ganoderma*	—

Ornamental mushroom

1.	*Microporous* sp.	—
2.	*Trametes* sp.	Turkey tail

**Table 3: Temperature Required for Storage and Incubation of
Important Edible Mushroom**

Sl.No.	Objects	Agaricus	Pleurotus	Lentinus	Volvariella	Calocybe
1.	Days for complete colonization of mother spawn	20-21	15-16	20-22	12-15	15-17
2.	Days for complete colonization in commercial spawn	12-14	12-14	15-16	12-15	12-14
3.	Incubation temperature (ºC) during colonization	25	25	25	32	28
4.	Storage temperature (ºC)	4	4	4	10-15	15
5.	Shelf-life of spawn	Two months	One month	Three months	< 15 days	15 days

Table 4: Cultivation Environment and its Cost Benefit Analysis of Important Mushroom

Sl.No.	Objects	Agaricus	Pleurotus	Volvariella	Calocybe
1.	Time	October-February	September-April	Last week of April to September	March to September
2.	Substrate	Wheat straw	Wheat straw, Rice straw	Rice straw	Wheat straw
3.	Optimum temperature for fruiting (ºC)	18-20	20-28	30-32	30-35
4.	Periods (days)	25-30	20-30	15-20	25-30
5.	Months	1-2	3-4	4-5	2-3
6.	Production	15-20 per cent	50-80 per cent	15-20 per cent	40-50 per cent
7.	Input/kg	₹ 40/-	₹ 30/-	₹ 30/-	₹ 30/-
8.	Sale/kg	₹ 150/-	₹ 120/-	₹ 130/-	₹ 150/-
9.	Benefit	₹ 110/-	₹ 90/-	₹ 100/-	₹ 120/-

Index

A

Agaricus bisporus 13-17, 19-22, 31, 68, 139, 155

Agro-climatic requirements 68, 80, 94, 97

Agro-waste 21

Amanita muscaria 15, 101, 102, 155

Anaemia 20

Annuls 27

Anti-metabolites 31

Antioxidant property 20

Anti-ulcer properties 20

Auricularia sp. 14, 89

Autoclave 2

Auxotrophs 31

B

Bavistin 12, 68, 75, 80, 84, 95, 154

Biochemical compounds 3

Biological oxygen demand (BOD) incubator 5

Bioremediation 23

Biscuit 115

Breeding programme 29, 30, 31

Breeding technique 31

Broadcast spawning 64

Button mushroom 14, 15, 17, 64, 67-71, 112

C

Calcium carbonate 12

Calocybe indica 17

Canning 106

Casing 57-62, 68-71, 80, 81, 145

Cecid flies 123

Cell-culture dish 10

Cereal straw 57

Chemical sterilization 12

Chemopreventive 20

Chinese mushroom 14, 83

Chips 114

Chloropicrin 61

Cleaning chemicals 2

Cob web 118

Commercial/planting spawns production 52

Compost 57, 59, 61, 64, 65, 68-70, 123, 146, 147

Controlled atmosphere packaging (CAP) 71, 108

Cork borer 2, 41

Cropping 77

Cropping room 2

Cryogenic freezing 42

Cultivation technique 67, 71, 73, 75, 77, 79, 80-99

Culture media 37, 39, 41

Culture tubes 2, 41

Cyanocobalamin 19

D

Deep freezer 7

Design 139

Dhingree mushroom 14

Diabetic patients 20

Dietary supplements 21

Dimethyl sulfoxide (DMSO) 42

Direct marketing 136

Diseases 117

Disorders 117

Distillation plant 8

Diversification 128

Dry bubble 118

Dry heat sterilization 12

Drying 107

Dudhiya mushroom 14

E

Edible mushrooms 13, 19, 89, 91

Entrepreneurship 127

Equipments 1-3, 5, 7, 9, 11, 153

Ethics 127

Export 133

F

Financial services 133

First phase 58

Flake spawn 49

Flame sterilization 12

Flammulina velutipes 22, 89, 90, 155

Flock 120

Fluorescent light 5

Formalin 12, 61, 64, 65, 68-70, 75, 80, 81, 84, 95, 97, 118, 120, 121, 123, 154

Free-drying 41

Freezing methods 41

Fruiting 70, 71, 76, 86, 96, 98, 145, 146, 161

G

Ganoderma lucidum 13, 20, 22, 36, 93, 96, 97, 98, 131, 155

Gills 26

Grain spawn 49

Grain substrate 50

Granular spawn 49, 146

Green mould 118

Growth environment 141

H

Hallucinogenic 101

Harvesting 142

Heterokaryotic 31

High blood pressure 20

High efficiency particular air (HEPA) filter 5

Hollow core 120

Hormone stimulator 21

Horse spawn 49

Hot air oven 3, 4

Hot plate 9

House management 139

Humid conditions 79

Humidifier 11

Humidity 29

Hybridization 31, 146

Hygrometer 10

Hypertension 20

I

Immune System Enhancer 21
Incubation Chamber 1
Ink caps 120
Inoculation bags 95

K

Karyogamy 30
Ketch-up 113
Khumb mushroom 14

L

Laboratory condition 138
Laboratory practices 1, 3, 5, 7, 9, 11
La France 119
Lambert's agar medium 38
Laminar flow cabinet 5
Layer spawning 63, 64
Lentinula edodes 14, 22, 93, 94
Life cycle 29, 30, 31
Log methods 89

M

Macro-fungus 25
Magnetic stirrer 9
Malt extract agar (MEA) medium 38
Manure spawn 49
Market channels 135
Marketing 133
Matured mushrooms 77, 90
Measuring cylinders 2
Medicinal importance 20
Medicinal mushrooms 13, 93, 95, 99, 155
Microbial activity 58
Milky mushroom 14, 79, 81, 82, 110, 112
Minerals value 20
Mites 123
Mixed spawning 63, 64, 65

Modified atmosphere packaging (MAP) 71, 106, 108
Moist heat sterilization 12
Moisture 59
Mother spawn production 50, 51
Multi-spore cultures 32, 41
Mummy disease 119
Murabba 114
Muscarine type 101
Mushroom candy 113
Mushroom cooking 111-115
Mushroom cultures 37, 39, 41
Mushroom house 2, 137, 139, 140
Mycelium 27
Mycorestoration 21

N

Natural or virgin spawn 49
Nuggets 115
Nutraceuticals 21
Nutrient medium 32
Nutrition facts 18

O

Organic manure 57, 69, 129, 130
Oyster mushroom 63, 64, 73, 75-77, 160, 161

P

Packaging 106-108
Paddy straw mushroom 14, 83, 85
Pasteurization 16, 58, 59, 75, 76, 80, 81, 85
Peak heat stage 58
Pests 117
Petri dish 3, 5, 10, 12, 32, 40, 41, 43, 44, 147, 153
Petri plates 40
Phalloidin type 101
Phorid fly 123

Pileus 26
Pin head initiation 70
Plasmogamy 30
Plastic bag methods 90
Poisonous mushroom 13, 27, 101
Polyethylene glycol (PEG) 31
Poly propylene bag 50, 68, 75
Post harvest management 71, 77, 82, 87, 105, 107, 109, 144
Post-peak heat stage 59
Potassium meta bisulphite (KMS) solution 113
Potato dextrose agar (PDA) Medium 37, 47
Powder spawn 49
Precautions 53
Pre-peak heat stage 58
Preservation 37, 39, 41
Procurement of spawn 69, 75, 81, 85
Pure culture 39, 43, 45, 46, 47

R

Ready-to-eat mushroom curry (RTE) 115
Reishi mushroom 96
Relative humidity 10, 65, 68-70, 74, 77, 80-82, 84, 90, 94, 97, 108, 119

S

Sciarid fly 123
Sexual reproduction 30, 31
Shaker 8
Shiitake mushroom 93, 94
Short method 57
Silver ear mushroom 91
Single-spore cultures 30
Socio-economic status 127
Spatula 2, 73
Spawned casing 61

Spawn laboratory 1
Spawn production 12, 49-55, 160
Spent mushroom substrate (SMS) 21, 130
Spirit lamp 10
Spore culture 41
Spore print 34, 43
Springtails 123
Staining 8
Stalk 27
Sterile cork borer 41
Sterilization 2-4, 12, 45, 54, 61, 76, 90, 107
Sterilized bottles 51
Storage 37-41, 71, 106, 109, 144, 156
Strain improvement 29, 30, 31
Substrate used 139
Sugarcane bagasse 57, 69
Synthetic rubber gasket 4

T

Test tube 10
Thermometer 10
Tissue culture 40
Tobacco spawn 49
Transport 109
Tremella fuciformis 89, 91, 155

V

Value addition 111-115
Vegetative growth 29
Vegetative mycelium 49, 60, 148
Ventilation 139
Volva 27
Volvariella volvacea 14, 19-22, 27, 31, 34, 83-87, 131, 155

W

Weighing balance 8
Weight loss 21

Wet bubble 118

Wheat extract agar medium 39

Wheat grains 51, 52

White plaster 119

Winter mushroom 90

Wire mesh trays 4

Wood ear mushroom 89

Y

Yellow blotch 119

Yellow mould 119

www.ingramcontent.com/pod-product-compliance
Lightning Source LLC
Chambersburg PA
CBHW050518190326
41458CB00005B/1579